Water and Society VI

WITPRESS
WIT Press publishes leading books in Science and Technology.
Visit our website for the current list of titles.
www.witpress.com

WITeLibrary
Home of the Transactions of the Wessex Institute.
Papers published in this volume are archived in the WIT eLibrary in volume 251 of WIT
Transactions on Ecology and the Environment (ISSN 1743-3541).
The WIT eLibrary provides the international scientific community with immediate and
permanent access to individual papers presented at WIT conferences.
Visit the WIT eLibrary at www.witpress.com/elibrary.

Sixth International Conference on Water and Society

Water and Society 2021

Conference Chairman

S. Mambretti
Politecnico di Milano, Italy
Member of WIT Board of Directors

International Scientific Advisory Committee

Organised by

Wessex Institute, UK
Politecnico di Milano, Italy

Sponsored by

WIT Transactions on Ecology and the Environment
The International Journal of Environmental Impacts

WIT Transactions

Wessex Institute
Ashurst Lodge, Ashurst
Southampton SO40 7AA, UK

Senior Editors

Water and Society VI

Editor

S. Mambretti
Politecnico di Milano, Italy
Member of WIT Board of Directors

WITPRESS Southampton, Boston

Editor:

S. Mambretti
Politecnico di Milano, Italy
Member of WIT Board of Directors

Published by

WIT Press
Ashurst Lodge, Ashurst, Southampton, SO40 7AA, UK
Tel: 44 (0) 238 029 3223; Fax: 44 (0) 238 029 2853
E-Mail: witpress@witpress.com
http://www.witpress.com

For USA, Canada and Mexico

Computational Mechanics International Inc
25 Bridge Street, Billerica, MA 01821, USA
Tel: 978 667 5841; Fax: 978 667 7582
E-Mail: infousa@witpress.com
http://www.witpress.com

British Library Cataloguing-in-Publication Data

A Catalogue record for this book is available
from the British Library

ISBN: 978-1-78466-423-7
eISBN: 978-1-78466-424-4
ISSN: 1746-448X (print)
ISSN: 1743-3541 (online)

The texts of the papers in this volume were set individually by the authors or under their supervision. Only minor corrections to the text may have been carried out by the publisher.

Preface

This Volume contains some of the papers presented at the 6th International Conference on Water and Society, held on line due to the COVID pandemic, organised by the Wessex Institute, UK and the Politecnico di Milano, Italy. The Conference was sponsored by the International Journal of Environmental Impacts and WIT Transactions on Ecology and the Environment.

The Water and Society Conference follows the success of the previous meetings, the first of which was held in Las Vegas in 2011, followed by the New Forest, UK, home to the Wessex Institute, in 2013, A Coruña, Spain, in 2015, Seville, Spain, in 2017 and in Valencia, Spain, in 2019.

Water is essential for sustaining life on our planet and its uneven distribution is a source of permanent conflicts. This issue, if not correctly addressed, will lead to realignments in international politics and the emergence of new centres of power in the world.

The role of society and its involvement with water is paramount. Over the centuries, civilisations have relied on the availability of clean and inexpensive water. This can no longer be taken for granted as the need for water continues to increase due to the pressure from growing global population demanding higher living standards. To meet the future demands for water, new standards, new training and additional support roles will best be delivered by those knowledgeable of the new technologies and direction of the industry.

Many technologically feasible solutions, such as desalination or pumping systems are energy demanding but, as costs rise, the techniques currently developed may need to be re-assessed. The Conference addressed the interaction between water and energy systems. Many pipelines have now reached the end of their life; new (fast and trenchless) technologies for pipe rehabilitation have been developed, but are rarely used in many countries, because of the limited knowledge of the appropriate methodologies.

This meeting also encouraged trans-disciplinary communication on issues related to the nature of water, and its use and exploitation by society. Reuse is one of the methodologies which have to be established in the near future, but which requires the social acceptation. The motivation for the conference was the need to bridge the gap between the broad spectrum of social political sciences and humanistic disciplines and specialists in physical sciences, biology, environmental sciences and health, among others. Policy makers need to be educated and advised on developing policies and regulations that will support the water systems of tomorrow.

The intention of the Water and Society series is to review these issues, as well as the more technical aspects of water resources management and quality, to help put forward policies and legislation that will lead to improved solutions for all. The international nature of the attendees ensures that the conference findings and conclusions enjoy a wide and rapid dissemination amongst the water related science and policy communities.

The papers selected for presentation and included in the Conference Proceedings are permanently stored in the WIT eLibrary as Transactions of the Wessex Institute (see http://www.witpress.com/elibrary).

The Editor
2021

Contents

Deconstructing power dynamics and prevailing discourses in hydropolitics:
The case of the Sixaola river basin
Fatine Ezbakhe, Christian Bréthaut, Tania Rodríguez-Echevarría
& Diego Jara 1

Further development of the Ebbsfleet water management system
dynamics model: Adjusting representation of processes and system
boundaries, incentivising stakeholder re-engagement, and exploring
potential for university teaching
Vladimir Krivtsov, Alessandro Pagano, Sangaralingam Ahilan,
Emily O'Donnell & Irene Pluchinotta 11

How management efforts of a large hydropower firm impact sustainability
in the Colombian Andes: A multivariate analysis of society's perceptions
Daniel Cortés-Borda, Jorge-Andrés Polanco & Manuela Escobar-Sierra 23

Quality of greywater in Oman and its treatment using a sustainable system
Mohammed F. M. Abushammala, Wajeeha A. Qazi
& Mohammed Fahad Abdul Latif 37

Degradation of biodegradable single-use plates and waste bags in
terrestrial and marine environments
Kaire Torn, Georg Martin & Greta Reisalu 49

Forest management to mitigate disasters caused by heavy rain
Koji Tamai 57

How social are flood risk management plans in Spain?
Guadalupe Ortiz, Pablo Aznar-Crespo & Ángela Olcina-Sala 65

Design of a sewage and wastewater treatment system for pollution
mitigation in El Rosario, El Empalme, Ecuador
Bethy Merchan, Paula Ullauri, Fernando Amaya, Lenin Dender,
Paul Carrión & Edgar Berrezueta 77

Reduction of carbon emissions in a Mediterranean urban wastewater treatment plant
Manuela Moreira Da Silva, Luís Cristovão, Duarte Marinho,
Eduardo Esteves, Gil Fraqueza & António Martins 87

Assessment of physicochemical and bacteriological parameters in the surface water of the Juan Diaz River, Panama
Quiriatjaryn M. Ortega-Samaniego, Inmaculada Romero,
María Paches, Arturo Dominici & Andres Fraíz 95

Author index 105

DECONSTRUCTING POWER DYNAMICS AND PREVAILING DISCOURSES IN HYDROPOLITICS: THE CASE OF THE SIXAOLA RIVER BASIN

FATINE EZBAKHE[1], CHRISTIAN BRÉTHAUT[1], TANIA RODRÍGUEZ-ECHEVARRÍA[2] & DIEGO JARA[3]
[1]University of Geneva, Institute for Environmental Sciences, Switzerland
[2]University of Costa Rica, Centro de Investigación y Estudios Políticos, Costa Rica
[3]Environmental Law Center, International Union for the Conservation of Nature, Germany

ABSTRACT

The Sixaola river basin, shared between Costa Rica and Panama, is generally seen as one of the basins most prone to transboundary water cooperation. In the spotlight of the international community since the 1990s, the Sixaola basin has been framed as a prime example for multi-actor, multi-level, and multi-sector transboundary water governance, especially with the creation of the Binational Commission for the Sixaola River Basin (CBCRS for its name in Spanish). Still, one key question remains: who (and how) is behind such conceptual framing of the Sixaola basin? In other words, what actors have succeeded in defining the social reality of the Sixaola, and how has this reality affected the water governance? By looking at the discourses' evolution in the Sixaola basin, we aim to deconstruct the power dynamics underlining such discursive struggle. We employ Marteen Hajer's argumentative approach to discourse analysis to identify dominant storylines, practices and discourse coalitions, emphasising the role of international organisations. With this discursive approach, we examine how the Sixaola problem's boundaries (and ideas guiding the solutions to such problem) have changed in the past three decades. Our preliminary results show that international organisations have successfully framed the Sixaola "*problematique*" as local, while infusing it with a constant regional Central American perspective. Moreover, transboundary water governance has focused on sustainable economic practices to prevent biodiversity loss and land/water degradation, but with large banana companies at the margins. Finally, it remains unclear how international organisations have translated their discourses into national policies and regulations, beyond creating and strengthening the CBCRS as a platform for transboundary water cooperation.
Keywords: transboundary waters, water conflict and cooperation, discursive hydropolitics, Sixaola.

1 SETTING THE SCENE: WHY TRANSBOUNDARY WATERS

Transboundary waters are ubiquitous. 150 of the 196 countries worldwide share transboundary waters [1] with the 310 transboundary river basins covering 47% of the world's land surface and serving 52% of the world's population. In 52 countries – such as Afghanistan, Austria, Niger, and Paraguay – transboundary river basins cover more than 90% of countries' total surface [1]. Considering such numbers, it is no wonder that the UN Secretary-General Antonio Guterres, during the 2017 UN Security Council, stated that "it is essential that nations cooperate to ensure that water is shared equitably and used sustainably" and called for countries to "commit to investing in water security to ensure durable peace and security for all communities and nations". Indeed, the very nature of transboundary waters adds another layer of complexity when it comes to management. It requires more coordination – and collaboration – across countries, sectors, stakeholders, and scales, since mismanaging these waters raises a host of adverse effects, from water scarcity to poor water quality, environmental degradation, and economic losses to the communities relying on them.

To minimise the risk of mismanagement, the last decades have witnessed an emergence of rules and principles pushing transboundary water cooperation. From global conventions – such as the 1992 UNECE Water Convention and the 1997 UN Watercourses

WIT Transactions on Ecology and the Environment, Vol 251, © 2021 WIT Press
www.witpress.com, ISSN 1743-3541 (on-line)
doi:10.2495/WS210011

Convention – to bilateral and multilateral treaties, agreements, and arrangements, transboundary water cooperation has been put forward as a means to reconcile uses and reduce tensions that come from the use of such waters. The Global Environment Facility (GEF) alone has invested nearly 10 billion in grants and co-financing to strengthen cooperation on international and transboundary basins, the most recent one on institutionalising transboundary water management in the Panj river sub-basin, between Tajikistan and Afghanistan.

The last decades have also brought a shift in how we understand the actors in transboundary water governance. Rather than viewing transboundary waters as a purely States' affair, there has been an increasing realisation that their governance involves a myriad of non-state actors across a multitude of levels, both geographical and institutional. In other words, transboundary waters are a product of a fluid interaction between state and non-state actors, with the divide between them increasingly blurred due to the partnerships and multi-actor platforms. We now speak of *governmentality*, not government [2], with non-state actors as integral part of the processes and outcomes of governing transboundary waters.

Out of all the non-state actors involved in transboundary water governance, in this paper, we put the spotlight on international organisations (IOs), including intergovernmental organisations (IGOs), international financial institutions (IFIs), and international non-governmental organisations (INGOs). We aim to explore how these non-state actors shape the governance of transboundary waters. Instead of focusing on their resources, power, interests, and so on, we adopt a discursive lens to shed light on how IOs achieve to define the problems to be solved and propose the policy responses to those problems. In other words, by turning our attention to the IO's discourses, we can gain insight into how such actors participate in the construction of a basin's social reality through their projects, programmes, and initiatives.

We take the Sixaola river basin, shared between Costa Rica and Panama, as our case study. Based on Hajer's [3] argumentative approach to discourse analysis – that is, storylines, practices, and coalitions as our analytical variables – we examine how IOs have participated in the framing of the Sixaola problem's boundaries (and the ideas guiding the solutions to such problem) over the past three decades.

2   WHY THE SIXAOLA RIVER BASIN

There are 310 transboundary river basins, some of them notorious. The Nile, the Mekong, the Amazon, the Ganges-Brahmaputra-Meghna, the Tigris-Euphrates, the Rhine; such river basins have been the subject of many studies. The Sixaola river basin, on the other hand, is at the other end of the spectrum, with much less analyses devoted to it [4], [5]. Therefore, reflecting on why we choose the Sixaola – a basin no bigger than 3,000 km^2 – as our case study is pertinent.

The Sixaola river basin remains "one of the most peaceful and prone to cross-border cooperation area in Central America", according to the European Commission [6]. Indeed, the relations between the two riparian countries, Costa Rica and Panama, do not resemble others in the isthmus. For instance, the Costa Rica–Nicaragua dynamics over the San Juan river basin are much more complex, underlined by unresolved territorial disagreements. The Costa Rica–Panama border has been undisputed since the 1941 Echandi-Fernández treaty, with several bilateral treaties signed in the 1970s and 1980s, from free trade (1973) and border protection and surveillance (1975) agreements to border cooperation (1979) and the creation of the La Amistad International Park (1982). Still, such peaceful diplomatic relations did not happen out of the blue. The 1941 border agreement arose after an armed conflict in the 1920s, known as the Coto War, which resolution came with the intervention of the United

States and the involvement of U.S. banana companies (e.g., United Fruit Company and American Banana Company) operating in the region.

Yet, it is not the overall peaceful nature of the Costa Rica–Panama that makes the Sixaola river basin an intriguing case study for our research. Instead, it is its multi-dimensional nature, illustrated by its environmental, economic, and social value. First, this small basin is internationally recognised as a global biodiversity hotspot, both in terms of fauna and flora. Host of many endangered species receiving conservation priority, the Sixaola basin has attracted the attention of IOs dedicated to environmental conservation. Today, several protected areas cover the basin – the La Amistad International Park (PILA), Hitoy-Cerere Biological Reserve, Gandoca-Manzanillo National Wildlife Refuge, San San Pond Sack Wetlands and Palo Seco Protection Forest – accounting for more than half of the basin's surface. Second, the Sixaola basin is home to many indigenous and Afro-descendent communities, nearly 60% of the population. Despite this social value, the Sixaola is one of the most deprived regions in both countries, with human development indexes below the national averages. Third, the basin has an agriculture-based economy. Specifically, export banana production remains the central pillar of the basin's economy and employment source, despite the increasing efforts to diversify the agriculture to alternative crops (mainly cacao, pineapple, tubers and fruit trees), and notwithstanding the negative environmental consequences from the intensive consumption of agrochemicals. This multi-dimensional and somewhat complex nature makes the Sixaola basin an attractive case study.

What is more, due to this very multi-dimensional makeup, the Sixaola basin has become an ideal candidate for international cooperation. Deemed by some as the "laboratory for cooperation", the Sixaola has stirred the interest of different intergovernmental organisations, international NGOs, and cooperation agencies. Since the early 1990s, the history of the Sixaola is replete with initiatives and projects proposed, funded, or implemented by international organisations, mainly dedicated to environmental conservation and sustainable development. For instance, as recently as in 2019, the GEF has approved a concept note for an 18 million project on "transboundary integrated water resource management" in the Sixaola river basin.

If we understand transboundary water governance in the broad sense – that is, as a multi-actor, multi-level, and multi-dimensional process – the Sixaola constitutes an ideal candidate for exploring how IOs have shaped the basin's social reality and how this reality translate in the way it is government. As several IOs have framed the Sixaola as a "first step", a justification to enter in more complex basins, it is right to use it as an excellent first study for our research purposes.

3 OUR ANALYTICAL FRAMEWORK

While there are many different approaches to do discourse analysis, we use Hajer's argumentative analysis as our source of inspiration, which entails two key elements. First, we see discourse analysis not only in its linguistic dimension – i.e., what we say or write – but mainly in its performative and practical dimensions – i.e., what are the practices through which we produce and reproduce what we say. In other words, for us, discourse analysis is about meaning, not language, and about the dynamic processes of meaning production. Second, to undertake discourse analysis, we use three factors: storylines, practices, and discourse coalitions. Storylines are a "generative sort of narrative that allows actors to draw upon various discourse categories to give meaning to specific physical or social phenomena". Thus, storylines serve as easy representations of an issue and help actors make their arguments. Practices are "through which meaning is given to physical and social realities", how actors translate or communicate a particular storyline. Finally, discourse coalitions are the "ensemble of a set of storylines, the actors that utter these storylines, and the practices that conform to them". Actors in the same discourse coalitions might not know each other, but they use the same argument and set of practices.

With this approach, the questions to answer are thus: (1) what storylines IOs use as arguments to their actions in the Sixaola? (2) what practices they use to turn these storylines into reality? and (3) what coalitions emerge from these storylines and practices? We will not cover the last question in the present paper, though, as it remains the object of ongoing work. We base our analysis on secondary documentation on the initiatives, projects and strategies financed and implemented by IOs in the Sixaola, complemented with semi-structured interviews with key actors to triangulate the information.

4 THE DISCURSIVE STRUGGLE IN THE SIXAOLA: AN EVOLUTION
1941, with the signing of the Border Agreement between Costa Rica and Panama, marks what both parties describe as a period of "friendship, cooperation, good neighbourliness, and peace" [1]. Yet, if one pays a closer look at what has happened since 1941 in the Sixaola, particularly from the discursive lens, three phases stand out: (1) Phase 1 (1970s–1995), with a focus on regional integration and peace; (2) Phase 2 (1995–2005), with an emphasis on biodiversity conservation; and (3) Phase 3 (2005–2015), with a momentum for local water governance. While the three phases are distinct, the start and end dates are not fixed but rather an interpretative and simplified depiction. In reality, the phases – and the discourses underlying them – overlap.

4.1 Phase 1 (1970s–1995): Towards regional integration and peace

The 1970s–1990s decades were ones of border agreements, which went beyond the Sixaola, and focused on peace and cooperation. The peace discourse is crystal clear in the first border agreement signed: the 1979 Agreement on Border Cooperation. While the 1979 Agreement puts border cooperation as a vital ingredient for socio-economic development, it also puts such cooperation in the context of the "fraternal ties between the countries". The 1979 Agreement thus links socio-economic development in the border with overall peaceful interactions between the two countries and aims to "adopt the pertinent measure to improve the region in all aspects and to regularize commercial exchange for mutual benefit". The same year, the two Governments signed the 1979 Joint declaration on the creation of the La Amistad International Park (PILA). The discourse framing the creation of this park is also in terms of peace: not only is the park explicitly labelled as a "friendship park", but more importantly, the park becomes a reflection of the "excellent relationships of friendship and brotherhood between the two peoples and governments", a way to "continue cooperation in the border area" and "conserve and preserve flora and fauna of the region". Under this discourse of "peace" and "friendship", the two countries then signed the 1992 Agreement on Cooperation for Border Development, which exists till this day. As with the previous two, this agreement builds on the "fraternal ties of friendship, mutual understanding and democratic vocation", with the aim of "broadening, improving and deepening the cooperative relations in all fields", contributing to the development in its different dimensions, i.e., "social, commercial, environmental and political improvement". At first glance, these three agreements seem to be a State-driven initiative, an endeavour between two States based on a discourse of peace and friendship. However, such agreements do not happen in a vacuum but are thoroughly situated in a regional context.

By looking at the Central American regional context, we can uncover how this discourse of peace and friendship is attached to a broader one, shaped mainly by IOs. Specifically, all three previous agreements stemmed from IOs-driven initiatives at the regional level. The 1979 Agreement on Border Cooperation, for instance, could not have been possible without the role of the Inter-American Development Bank (IDB), a multilateral financial institution established in 1959 to finance "economic and social development" in Latin America and the Caribbean. The 1979 Agreement resulted from a 1971 Study for Integrated Development of the Border Region, funded with the IDB's the Pre-investment Fund for the Integration of Latin America created in 1966. The discourse of regional integration is very evident in the Fund's aim, which targets "studies that enable the identification and preparation of multinational projects in all areas that are important to promote regional integration". While the two Governments never completed the 1971 IDB-funded study, they used it as the basis for developing the 1979 Agreement. IOs were also involved in the shaping of the 1979 Joint Declaration of the creation of the PILA. Five years before the declaration, the International Union for the Conservation of Nature (IUCN) organized the First Central American Meeting on Natural and Cultural Resource Management. The 1974 meeting brought together governments of the countries and experts from concern IOs (UNEP, FAO, UNESCO, WWF, CATIE) to "develop an action programme for an integrated regional system of national parks and equivalent reserves". One of the seven recommendations resulting from the discussions put the spotlight on "border areas that are particularly well suited to be frontier parks between neighbouring countries". Furthermore, we cannot understand the 1992 Agreement on Cooperation for Border Development without considering the role of the Central American Integration System (SICA). This economic and political organization, created in 1991, embodied "a new vision of Central America as a region of peace, democracy, and development", a way to overcome the turmoil and violence witnessed in the region in the 1980s. In 1992, the "Foro de Vicepresidentes Centroamericanos", a body of the SICA, emphasized the border development in the region and supported the elaboration of a regional action plan for "border development and integration". Thus, it is not surprising that the discourse of peace and friendship found in the 1992 Agreement is similar to other agreements signed in this period, such as the 1992 Trifinio Plan and the 1993 Treaty for the Execution of the Trifinio Plan.

Thus, in this first phase, we can identify three distinct storylines:

- Storyline 1: Regional integration as a "requisite" for sustainable peace and cooperation. Regional integration is both a goal and a means to achieve long-term and sustainable peace and cooperation.
- Storyline 2: Border regions as "ideal" spaces for cooperation and development. Border areas are no longer seen as conflict-prone areas to avoid but as the perfect spaces for strengthening cooperation. Under this rationale, border areas are the first step to achieving regional integration through frontier parks.
- Storyline 3: Development from a socio-economic perspective. Peace and development are tools for socio-economic improvement. Even frontier parks are not about protecting the environment per se, but about improving the wellbeing of the people living in these border areas.

In this same phase, we can identify three concrete practices that IOs used to utter the storylines:

- Practice 1: Institution creation. The most important institution created was the SICA, aimed at regional integration, but IOs also supported the establishment of binational and trinational governance structures.

- Practice 2: Agreements and treaties. The agreements and treaties signed in the early 1990s followed the discourse on regional integration and peace.
- Practice 3: High-level meetings. In particular, the meeting organised by IUCN translated into the creation of binational parks.

Finally, in terms of discourse coalitions, all IOs followed similar discourses, without any opposing storylines.

4.2   Phase 2 (1995–2005): Emphasis on biodiversity conservation

Furthermore, the Sixaola soon became a piece of the regional biodiversity conservation puzzle. Since the 1995s, the Sixaola became host of a series of regional environmental projects – including the Parks in Peril, Regional Environmental and Natural Resources Management (RENARM), Regional Environmental Program for Central America (PROARCA), and the Mesoamerican Biological Corridor (MBC) – funded by IOs. All these projects followed a similar discourse, one of "minimising damage to the environment, protecting biodiversity, and providing the means for equitable and sustainable economic growth" and of "creating local capacity for the conservation of threatened, high-biodiversity landscapes". Rather than the multi-dimensional approach promoted by the Binational Governance bodies, this string of projects emphasised the environment. Plus, these projects introduced a new discursive element: a need for creating local capacities, not for decision-making but somewhat operational, for the conservation of biodiversity landscape. However, these environmental projects came in juxtaposition with the Plan Puebla Panama, a regional investment plan launched in 2000 to promote "productive investment to boost economic development and overcome poverty" and attract "productive private investment to the region". In the 2000–2006 period, the Plan Puebla Panama funded more than 33 megaprojects, totalling USD 4,500 million, most of them for transportation and energy infrastructure. Some scholars describe it as a "neoliberal vision of development" [7], [8], contradicting the biodiversity conservation discourse promoted by IOs.

In this second phase, the storylines of IOs evolve into:

- Storyline 1: Biodiversity of the region as a "global ecological issue". In a sense, biodiversity loss in Central America concerns the whole world, making it a priority of IOs.
- Storyline 2: Environmental conservation and protection as a goal. The preservation and protection of ecosystems is, in itself, a goal to achieve, for the sake of the environment.
- Storyline 3: Environmental conservation and protection as a means. Natural resources are assets with socio-economic value and thus need to be conserved because to "boost" socio-economic development.

In terms of practices of the IOs, we can distinguish two main ones:

- Practice 1: Development cooperation projects. INGOs and States' foreign aid agencies employed projects to make environmental conservation a reality in the Sixaola.
- Practice: Investment strategies and funds. Regional institutions and banks employed regional investment strategies, most importantly the Plan Puebla Panama, to put environmental conservation part of the socio-economic development of the region.

In this sense, two discourse coalitions emerge. On the one hand, there is the coalition of INGOs, where biodiversity conservation remains a goal, that is, conserving ecosystems for the sake of ecosystems. On the other hand, the coalition of most regional banks and institutions frames biodiversity conservation as means to achieve socio-economic development.

4.3 Phase 3 (2005–2015): A momentum for local governance.

The third phase kicks off with the two national projects for the Sixaola: the *Program of sustainable development of the Sixaola river binational basin* in Talamanca (Costa Rica) and the *Multiphase program of sustainable development in Bocas del Toro* in Bocas del Toro (Panama). Both projects, each in one part of the basin, emerged from the Executive Secretariats' efforts of the *1992 Agreement on Cooperation for Border Development.* They were "interventions in the economic, social, and environmental spheres" to contribute to a "sustainable development model". While the discourse on the three pillars of sustainable development is not new, the two projects added a new element to the Sixaola's discursive evolution: the need for pilot projects aimed at productive diversification. Specifically, the projects promoted organic cacao as an alternative to banana monocultures, but exclusively in small farms, leaving the large banana companies outside of the projects' scope. In 2006, the Executive Secretariats went one step further by approaching the Global Environment Facility (GEF) for funding a binational project, covering the basin as a whole. The process of developing this binational project framed the "Sixaola problem" as one of "limited sustainable livelihoods, unsustainable economic activities, and institutional limitations" and the "Sixaola solution" being institutional strengthening, but without the creation of new institutions. However, this discourse shifted when the GEF-funded *Binational Sixaola Project on Integrated Management of Ecosystems* started in 2008. The project required the creation of a binational commission to provide "strategic direction" to the project and be responsible for its supervision. The reasons behind this discursive change remains unclear but can be found in other GEF-funded projects, where local commissions are requirements to foster participation. Due to this requirement, the countries created the *Binational Commission of the Sixaola River Basin* (CBCRS) in 2009, with another spin in the discourse: first, it was not a requirement, but a choice by the actors involved to have a body under the umbrella of the 1992 Agreement; second, it was no longer for project supervision, but for a "long-term vision of becoming a unit for local planning and management of the basin". Furthermore, as in the previous phase, the Sixaola was host to a series of regional projects, including Alliances, Water Management for Adaptation, Adaptation, Vulnerability and Ecosystems (AVE), and Building River Dialogue and Governance (BRIDGE), all implemented by IUCN. All these projects followed the same discourse: that local capacity building is critical for facilitating transboundary water cooperation since there are more opportunities for change at the local level. In the Sixaola, IUCN-projects targeted the strengthening of the Binational Commission, aiming to make it a "platform" for multi-actor (including indigenous communities) and multi-sectoral (including tourism and economic practices) decision-making. Finally, in 2010, the adoption of the "Central American Strategy for Rural Development" (ECADERT) represented a different shift in the discourse. First, the focus was no longer on border areas but rural territories (although the strategy included border areas such as the Sixaola because of their rural and peripheral nature). Second, there was an emphasis on public policies instead of projects. Third, there was an explicit promotion of organic cacao production to achieve "inclusive economies for rural territories".

The storylines emerging from these projects and strategies are:

- Storyline 1: Local governance as key to inclusive transboundary water governance. The rationale is that, since governance happens at the local level – both formal and informally

– it is necessary to bring together all the different actors and sectors and foster dialogue between them.

- Storyline 2: Unsustainable economic activities as a root cause. Issues such as biodiversity loss and water degradation cannot be separated from the unsustainable agricultural practices in the basin. Still, the storyline does not shed light on all the actors behind such agricultural practices (large banana companies in particular).
- Storyline 3: Institutional strengthening as a requirement for sustainable development. Institutions – at the local level, in particular – lack human and financial capacities, and thus strengthening them becomes the first and must-do step towards achieving sustainable development.

Again, there are two distinct practices that IOs used to utter these storylines:

- Practice 1: Development cooperation projects. Most of these projects focused on the strengthening of the Binational Commission.
- Practice 2: Direct advice, support, and technical assistance.

As in the first phase, there are no opposing discourse coalitions. All IOs subscribe to the discourse of local water governance as both a goal and means for the basin's sustainable management.

5 THE DISCURSIVE STRUGGLE IN THE SIXAOLA: AN EVOLUTION

In this paper, we analysed the discursive struggle in the Sixaola river basin, particularly the role of IOs in the boundaries and framings of the problem and the ideas for the solutions proposed. From our analysis, we can highlight the following aspects:

- Boundaries. The regional perspective has been a constant in all discursive phases. The "Sixaola problem" is not only bounded to the Sixaola basin, but linked to other basins in Central America. In the last phase, we see an increasing focus on the local level, especially for inclusive governance processes.
- Conceptual framings. The root cause of the "Sixaola problem" lies in the unsustainable economic practices, mainly in the agricultural sector, and hence the efforts towards organic cacao. Yet, throughout the years, the discourse of IOs miss a critical piece of the puzzle: the large banana companies present in the basin.
- Ideas guiding the "Sixaola solutions". What started as signing treaties and agreements slowly evolved into multi-actor platforms, now focusing exclusively on the Binational Commission. Still, this operationalisation of the solutions has always focused on projects instead of public policies. This reflects that the discursive power of IOs has been through the formulation and implementation of projects and leaves the question of why not public policies.

ACKNOWLEDGEMENTS

We thank the interviewees who generously gave their time in order to contribute to this research. This research is part of a broader project on *Monitoring for International Hydropolitical Tensions*, jointly developed by the University of Geneva, Oregon State University, Tufts University, and the International Union for the Conservation of Nature.

REFERENCES

[1] McCracken, M. & Wolf, A.T., Updating the register of international river basins of the world. *International Journal of Water Resources Development*, **35**(5), pp. 732–782, 2019.

[2]   Foucault, M., Governability. *The Foucault Effect: Studies in Governmentality,* eds G. Burchell, C. Gordon & P. Miller,. The University of Chicago Press: Chicago, USA, pp. 87–104, 1991.

[3]   Hajer, M.A., *The Politics of Environmental Discourse: Ecological Modernization and the Policy Process*, Oxford University Press, 1995.

[4]   Rodríguez-Echevarría, T., Gobernanza ambiental en cuencas transfronterizas: la cuenca del río Sixaola (Costa Rica-Panamá), *Iztapalapa Revista de Ciencias Sociales y Humanidades*, **87**(40), pp. 71–99, 2019.

[5]   Rodríguez-Echevarría, T., Obando, A. & Acuña, M., Entender el extractivismo en regiones fronterizas: Monoculuvos y despojo en las fronteras de Costa Rica. *Revista Sociedad y Ambiente*, **6**(17), 2018.

[6]   European Commission, EU – Central America Cooperation: Support to the Central American Integration System's (SICA) cross-border cooperation actions. Final Report. Association of European Border Regions (AEBR), 2014.

[7]   Cadena-Montenegro, J.L., El Plan Puebla Panamá: ¿La reconolización de América Latina? *Revista de Relaciones Internacionales, Estrategia y Seguridad*, **1**(2), pp. 121–155, 2006.

[8]   Ken-Rodríguez, C.A., La iniciativa de integración del Plan Puebla Panamá: consideraciones desde la postura del desarrollo regional equilibrado de Mesoamérica, *Economía y Administración*, **2**(2), pp. 95–126, 2017.

FURTHER DEVELOPMENT OF THE EBBSFLEET WATER MANAGEMENT SYSTEM DYNAMICS MODEL: ADJUSTING REPRESENTATION OF PROCESSES AND SYSTEM BOUNDARIES, INCENTIVISING STAKEHOLDER RE-ENGAGEMENT, AND EXPLORING POTENTIAL FOR UNIVERSITY TEACHING

VLADIMIR KRIVTSOV[1,2,3], ALESSANDRO PAGANO[4], SANGARALINGAM AHILAN[5],
EMILY O'DONNELL[1] & IRENE PLUCHINOTTA[6]
[1]University of Nottingham, UK
[2]RBGE, UK
[3]University of Edinburgh, UK
[4]Water Research Institute, National Research Council, Italy
[5]University of Exeter, UK
[6]University College London, UK

ABSTRACT

Sustainable urban water management must address interconnected social, technical and environmental issues. Modelling helps us understand these interconnections and provides a tool to analyse interactions between the urban water system and alternative management strategies. Models may be used to simulate not only the effects of climate, social and economic changes, but also the impacts of technological innovations, and different policy interventions. This paper reports an update of the Ebbsfleet Water Management System Dynamics Model (SDM) that explores sustainable urban water management. The model is open-source compiled using Vensim software, which is free for non-commercial use. This paper demonstrates that the current SDM and the modelling approach are open to adjustment, which is illustrated by introducing a link between water tariffs and environmental awareness. The increase in water tariffs leads not only to the obvious increase in water bills, but also to an increase in Environmental Awareness, and consequently, to increases in the use of water efficiency devices, grey water acceptability, and grey water reuse. A range of further modifications is suggested, including expanding representation of sustainable drainage systems (SuDS) to consider resultant improvements in stormwater quality as well as quantity. This would recognise the indirect benefits of improved stormwater quality on biodiversity in the River Ebbsfleet, which is the receiving watercourse. This study intends to encourage knowledge transfer, by facilitating and incentivising the use and further development of the SDM by stakeholders and a wider community of end-users, including practitioners, academics and the public. While SDM is particularly suited to analysis of indirect relations, benefits and trade-offs among system constituents, other approaches provide viable alternatives and we discuss the potential for re-implementing our findings in other interactive modelling software packages and programming languages. We also explore the scope for linking the adapted SDM to other models. Finally, we consider the utility of the Ebbsfleet SDM in teaching, learning and knowledge transfer. We conclude that students, practitioners and other stakeholders could not only enhance their understanding of urban water management complexity, but also gain valuable system modelling skills based on using the SDM to support kinaesthetic learning. Ultimately, society benefits when the level of knowledge and analytical thinking skills of its members are enhanced.
Keywords: water balance, system thinking, water security, societal benefits, sensitivity analysis, nature-based solutions, industrial ecology, public engagement, pricing policy, 'what-if' scenarios.

1 INTRODUCTION

Urban water and flood risk management are increasingly regarded as essential components in sustainable and healthy living as well as economic activities in cities [1]–[3]. Modern

WIT Transactions on Ecology and the Environment, Vol 251, © 2021 WIT Press
www.witpress.com, ISSN 1743-3541 (on-line)
doi:10.2495/WS210021

urban water management is a very complex process associated with a multitude of interconnected social, technical and environmental issues [4]. A growing number of issues is expected to emerge concerning sustainable water resources management. Simulation modelling has been historically used for increasing the understanding of dynamically evolving urban water systems as it helps to elucidate this complexity and provides a potential tool for the analysis of interactions between system components. It also enables the assessment of a wide range of urban water management strategies related to technological innovations, policy interventions, climate change and meteorological variability, and social and economic issues [5], [6].

Three types of simulation modelling approaches are commonly used to explore urban water management and governance: urban water metabolism [7], [8], agent-based [9] and system dynamics modelling [10]–[12]. The metabolism-based modelling approach overcomes issues commonly encountered by independent modelling of the components of the urban water system (water supply, wastewater and surface water collection) by providing an integrated approach that considers the interconnection and interdependencies between water flows and other fluxes in urban systems including wastewater, energy and material. This integrated modelling approach enables the urban water system to be modelled over long-term planning horizons for both business as usual and future intervention strategies such as rainwater harvesting, greywater recycling, and wastewater reuse. While urban water metabolism provides a holistic picture of urban water management, there are several challenges and limitations in developing a detailed metabolic model e.g. data availability and uncertainty at high resolution [13]. Agent-Based Modelling has been used as well, but also with limitations e.g. in providing a comprehensive and cross-sectoral analysis of complex urban systems and in the analysis of future evolution through 'what-if' scenarios [9]. The System Dynamics Modelling approach is able to incorporate both the dynamic evolution of the urban system as well as social conditions such as stakeholders' priorities and goals. The multiple strengths of this approach have led to the development of many user-friendly interactive modelling tools such as STELLA, ModelMaker, Madonna, Simile and Vensim.

System dynamics modelling that investigates and subsequently resolves complex issues related to urban ecology and water management may be particularly fruitful when carried out together with stakeholders. The participatory process involving collaboration between a diverse community of practitioners and university researchers within the Ebbsfleet 'Learning and Action Alliance' (LAA) has been documented previously in comprehensive detail [14]. This paper aims to report on recent (i.e. after the publication of the initial paper) developments and further plans related to the Ebbsfleet System Dynamics Model (SDM). It therefore adds value in relation to the previous work, and has the following objectives:

- to demonstrate that the current SDM and the modelling approach are open to adjustment by modifying the model and suggesting further modifications;
- to demonstrate the capacity of the model to analyse indirect interrelations among system components;
- to identify the scope of linking the Ebbsfleet SDM to other models;
- to encourage knowledge transfer, by facilitating and incentivising the use and further developments of the SDM by stakeholders and a wider community of end-users, including practitioners, academics and the public;
- to identify and explore possibilities of using the Ebbsfleet SDM as a case study for teaching of simulation modelling, system dynamics, water security, and water management.

2 EBBSFLEET GARDEN CITY AND THE NEED FOR SDM

Ebbsfleet Garden City is situated on the bank of the River Thames and is part of the national government initiative to create a new Garden City in north Kent in response to the growing housing demand in South East England. The city aims to provide up to 15,000 good quality new homes built over twenty years (2015–2035), predominantly on former quarries, brownfield land and previously developed industrial sites. The development straddles the boundaries of Gravesham and Dartford districts and includes a number of strategic sites. The Ebbsfleet Development Corporation is a government organisation that actively works with landowners, developers, and water companies to successfully implement the development framework. In the Garden City, the water supply comes from both Thames (67%) and Southern Water (33%) and the wastewater management is largely dealt with by Southern Water. A growing population, water scarcity, prevailing complex landscape features, ageing infrastructure and involvement of multi-stakeholders in decision making highlight a need for a joined-up planning approach for sustainable integrated urban water management in Ebbsfleet. To meet this need, the system dynamics modelling approach has been used within the Urban Flood Resilience (UFR) project to explore sustainable water management in the Garden City [14]. As part of the participatory modelling process, a LAA was established, involving a range of key stakeholders (Thames Water, Southern Water, Ebbsfleet Development Corporation, Kent County Council, Dartford and Gravesham Borough Councils, among others). It should be noted that both the Ebbsfleet LAA and the SDM have been documented in comprehensive detail previously [14], and only a brief overview of the original model will be given in this paper. Further details, in particular the list of the equations, can be found in the previous publication whilst both the code and the compiled model are available from the authors.

3 EBBSFLEET SDM

The previously developed stock and flow model focused on one of the problem dimensions (sustainable urban water supply management) that emerged during the participatory exercises that were carried out in the Ebbsfleet Garden City. Based on a comprehensive conceptual system analysis (allowed also by the development of a set of Causal Loop Diagrams (CLD) in early stages of stakeholders' involvement) and complemented by relevant technical/scientific information (both provided by experts in Urban Water Management and available in the scientific literature), the model aims to explore, in a structured way, the evolution of the system over a 30 year time span and the potential effectiveness of multiple policy interventions. Indeed, one of the main advantages of using an SDM is the quantitative analysis of the impacts of manifold strategies that can be performed on the technical side (e.g. increase of the supply), on the socio-cultural side (e.g. educational programmes for water saving) or on the economic side (e.g. introduction of a water tariff). The model can be used to compare the effectiveness of different strategies as well as analysing their potential synergistic effects, mutual influences, unintended consequences and trade-offs.

The model (Fig. 1) computes a basic hydraulic water balance at the urban level (i.e. a comparison between water demand and water supply), performed at a yearly time step, and considers aggregated behaviour over the whole urban area (i.e. no individual or micro-behaviours are modelled). Specific attention, based on inputs provided by the stakeholders, is given to the analysis of the role of sustainable water saving/management strategies such as Rainwater Harvesting (RWH) and Grey Water Reuse (GWR). The model comprises stocks (i.e. accumulations related to real-world categories such as materials or knowledge, drawn with rectangles), flows (rates of change in the value of stocks, drawn with an arrow with a valve), variables (dynamic variables, used to define intermediate concepts and changing instantaneously according to an equation, and numeric constants) and links (arrows denoting a dependency between elements of a stock and flow diagram).

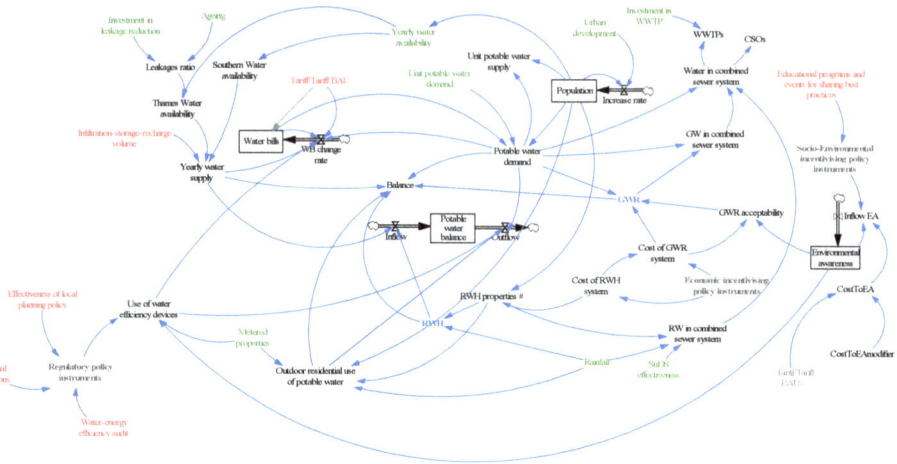

Figure 1:   Diagram of the Ebbsfleet SDM with an example of structural modification introducing a link between water tariffs and environmental awareness. Note that different colours are just a historic legacy (they simply refer to the stages at which the components were added during the model development).

It should be noted that the different colours in Fig. 1 are a historic legacy and refer to the stages at which the components were added during the model development. For example, the variables in green identify the first set of input variables selected with the stakeholders. The variables in red identify additional input variables added during the last workshop which aim to represent measures/policies capable of modifying the state of the system and its dynamic evolution. These colours are not essential for further development and users are encouraged to modify the colour scheme to represent their own preferences.

The representation of processes necessary to calculate the water balance in the Garden City have been detailed in the previous publication [14]. A brief overview is given below. The water supply depends on the 'availability' of water from both Thames Water Utilities Limited (for which the influence of 'leakages ratio' is explicitly considered) and Southern Water. The 'Infiltration-storage-recharge volume' is an additional volume that could theoretically be considered if specific actions are implemented. The global 'yearly water availability' is assumed to match the current per capita water consumption. The demand directly depends on the dynamics of the 'population' stock which follows the 'urban development' plan. A total of 12,000 houses (with average occupancy of three people) is foreseen by the end of the project. An interconnection between water supply and demand is explicitly considered, regulated by the 'water bill'. A baseline unit value (corresponding to the current water tariff) is considered, but an increase can be either activated under specific conditions (please refer to the model equations for full details) or limited by the 'use of water efficiency devices'. In the present work, a connection between the water bill and the 'environmental awareness' is also explored, as detailed in the next section. This connection reflects both the observational evidence and suggestions voiced by a minority of stakeholders during the LAA workshops.

In addition to the processes noted above, the water balance is affected by specific water saving measures (RWH and GWR), whose role is explicitly described in the model. The contribution of RWH systems is related to both the fraction of households with such systems installed out of the total number of properties ('RWH properties') and to the 'Rainfall', and contributes to the reduction of the 'outdoor residential use of potable water'. The uptake of RWH is considered to be directly dependent on the economic feasibility ('cost of RWH system'). The role of GWR is directly dependent on both the level of 'GWR acceptability' by the end-users (which is affected by the 'environmental awareness'), and on the 'Cost of GWR systems'. Both RWH and GWR, besides reducing the potable water demand, contribute to reduce the volumes of 'water in combined sewer systems'.

One of the key elements of the model is represented by the 'environmental awareness', which affects the 'GWR acceptability' as well as the 'use of water efficiency devices'. A direct influence of 'Socio-Environmental incentivising policy instruments' on the 'environmental awareness' is considered, as well as, for the purposes of the present work, on the water pricing policy. Several 'regulatory policy instruments' influence the 'use of water efficiency devices'.

The main output of the model is the 'Potable water balance', which is directly or indirectly affected by all the processes and model components mentioned above. It is worth highlighting that negative values of the potable water balance identify a potential water supply deficit. Conversely, positive values are mainly related to the assumption that the 'supply side' does not adapt with time e.g. to the reduction of potable water demand that would be associated to GWR and RWH or any other corrective action. In other words, the water supply follows an increasing water demand calculated according to the increase in population and a reference per capita water consumption. It should be noted that 'Potable water balance' is modelled as a stock. Its dynamics, therefore, result from differences between 'inflow' and 'outflow' and is described using a differential equation. For convenience, there is also a variable called 'Balance', which tracks the annual differences between the sum of replenishing processes (yearly water supply, RWH and GWR) and the sum of demands (i.e. potable water demand and the outdoor residential use of potable water).

4 LIMITATIONS OF THE PUBLISHED MODEL AND POTENTIAL FOR FURTHER DEVELOPMENTS

It is of course worth reiterating that any model is inevitably a simplification of reality, and any modelling effort is limited by time, logistical and financial constraints. Furthermore, the published SDM model [14] was designed in accordance with the outcome of discussions held during the LAA workshops, and is therefore constrained by the system boundaries and structural relationships thus imposed. It should, however, be noted that the current model structure is open to adjustment, and the users are indeed encouraged to explore not only developing further scenarios, but also making modifications to the model structure. Vensim software, which is sufficiently user-friendly and free for non-commercial use, was specifically chosen to enable further developments of the model by the end users.

4.1 Modification 1: Linking water tariffs and environmental awareness

One possible simple modification of the current model structure could be a link between water tariffs and environmental awareness (Fig. 1). The modified model diagram contains an explicit representation of the 'Tariff/Tariff BAU' parameter, i.e. the ratio between the water tariff used in any particular simulation and the baseline water tariff (BAU refers to Business As Usual). Although this parameter has previously been included in model definitions (see equation 38 in the original model), it was for simplicity omitted from the original model diagram as increase in tariffs was not investigated [14]. It should also be noted that 'Tariff/Tariff BAU' on the current diagram appears in two places; in addition to its original placement next to 'Water Bills' it is also shown on the right side of the diagram as a shadow variable.

The modified CLD also contains two completely new definitions: a parameter CostToEAmodifier and a variable CostToEA. The latter is described by the following equation:

$$CostToEA = 1 + ('Tariff/Tariff\ BAU' - 1) * CostToEAmodifier \qquad (1)$$

The equation 'Inflow EA' has also been changed to include multiplication by CostToEA. Depending on the value of the CostToEAmodifier parameter, the increase in water tariff will be transferred to the 'Inflow EA' using CostToEA variable either with no change (if CostToEAmodifier = 1), with reduction (if $0 \leq$ CostToEAmodifier < 1), or with amplification (if CostToEAmodifier > 1). When CostToEAmodifier = 0, the increase in tariff will not be transferred and the modified model becomes equivalent to the original model.

Fig. 2 presents the simulation results for scenarios detailed in Table 1. Scenarios T25up and T50up explore the influence of, respectively, 25% and 50% increase in water tariffs,

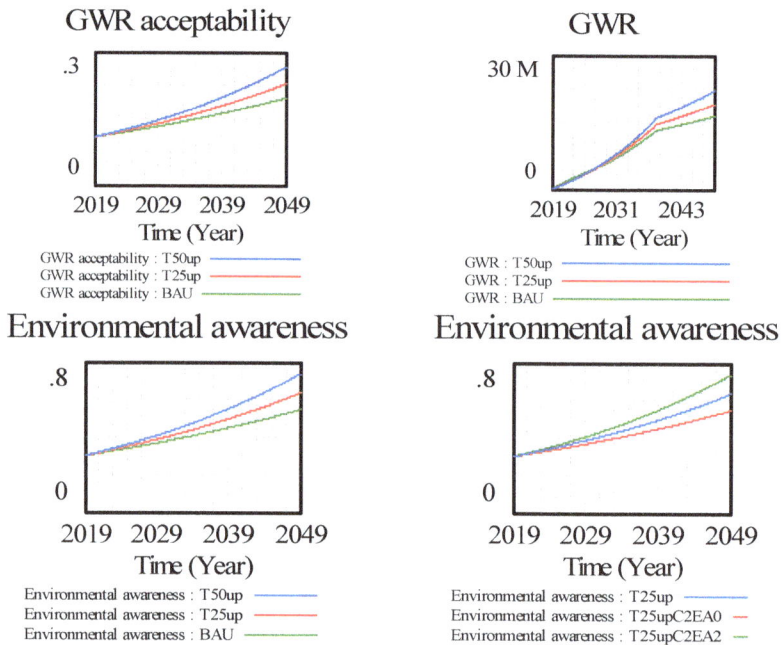

Figure 2: Examples of the simulation runs obtained using the modified model. The upper panels and lower left panel show the results of sensitivity analysis for the 'Grey Water Reuse' (GWR), 'Environmental Awareness' and 'GWR acceptability' variables on changes in the 'Tarif/Tarif BAU' parameter. The lower right panel presents the results of sensitivity analysis for 'Environmental Awareness' variable on the CostToEAmodifier parameter. The parameter values for these runs are given in Table 1.

Table 1: Parameter values used for specific scenarios.

Scenarios	CostToEA modifier	'Tariff/Tariff BAU'	Comments
BAU	1	1	This is the baseline scenario, but with an introduced link between the tariffs and 'Environmental Awareness'
T25up	1	1.25	This scenario simulates a 25% increase in water tariffs
T50up	1	1.5	This scenario simulates a 50% increase in water tariffs
T25upC2EA0	0	1.25	This scenario simulates a 25% increase in water tariffs but disables the link between the tariffs and 'Environmental Awareness'
T25upC2EA2	2	1.25	This scenario simulates a 25% increase in water tariffs and augments its influence on 'Environmental Awareness'
BAU0	0	1	With this combination the model simulations become equivalent to the baseline simulations of the original model

whilst scenarios T25upC2EA0 and T25upC2EA2 combine the 25% increase in water tariffs with changes in the value of the CostToEAmodifier parameter. The increase in water tariffs leads not only to the obvious increase in water bills (not shown), but also to the increase in Environmental Awareness, and consequently to the increases in the use of water efficiency devices, Grey Water acceptability, and Grey Water Reuse. However, there are no obvious changes in the dynamics of any other variables examined (data not shown).

It is also evident that changes in the value of the CostToEAmodifier parameter can be used to influence the link between Environmental Awareness and the water tariffs. For example, setting this parameter to 2 leads to the augmented increase in Environmental awareness, and consequently to the enhanced increases in the use of water efficiency devices (not shown), Grey Water acceptability, and GWR. However, setting the value of the CostToEAmodifier parameter to 0 disables the structural link between Environmental Awareness and the water tariffs, and the modified model becomes equivalent to the original model.

Interestingly, the influence of water tariffs on water bills is not simple, but in fact consists of the combination of direct and indirect effects. The former is straightforward (users pay more when on a more expensive tariff), whilst the latter relates to behavioural changes that lead to a decrease in potable water consumption due to an increase in environmental awareness and the usage of water efficiency devices.

4.2 Further modifications

Further possible changes to the model may relate both to the interconnections of the variables already in the model, and to the changes in system boundaries. It is hoped that advanced users would help to develop new versions of the model, and whilst doing so would benefit from achieving a better insight into the complexity of interlinked processes related to the sustainable management of urban water supply. Below are just a few suggestions of such potential modifications.

4.2.1 Modification 2

In the original SDM, grey water is assumed to be discarded through a combined sewer system. It would, however, be very easy to include a detailed representation of the stormwater pathways, and also include SuDS components, e.g. swales, green roofs and retention ponds.

4.2.2 Modification 3

Following the implementation of Modification 2, it would be possible to consider the runoff water quality and assess the combined effects of hydrology and water chemistry on the biological community of the SuDS retention ponds, as well as impact on downstream ecosystems. Such a representation of SuDS for stormwater treatment and consideration of runoff chemistry [15] are particularly interesting for the analysis of indirect effects [16], [17] related to the water quality and biodiversity of the receiving water courses [18]–[21].

4.2.3 Modification 4

An interesting possible development of the model may be a simulation of retrofitting green roofs. Such a scenario would have implications both for the quality of harvested rainwater, and for the quality and quantity of runoff. The combination of these changes is likely to lead to some complex interactions comprised of both direct and indirect effects, and simulation modelling is a very useful tool for untangling those and clarifying the emerging patterns.

4.2.4 Other modifications

The implementation of modifications 3 and 4 presents a further possibility to assess the impact on local wildlife, and to compare the contributions of various scenarios towards local biodiversity enhancement. Further useful modifications may include a more comprehensive representation of the population changes in Ebbsfleet (i.e. considering changeable rates of birth, death and migration), linkages between water supply and demand, and the incorporation of the rebound effect. The rebound effect (also known as the 'efficiency paradox') refers to a potential increase in water consumption due to water price reduction, which may occur when more water becomes available in the system (e.g. due to the increase in the system's efficiency). This is only partially considered in the model in its current form, since the reduction of water bills causes an increase of potable water use, but could be easily explored further in the future.

5 DISCUSSION AND CONCLUDING REMARKS

Simulation modelling is helpful for understanding the patterns of interactions between and within complex social and environmental systems. System dynamics modelling is particularly useful as it provides a convenient tool for the analysis of interactions between system components and the assessment of a wide range of potential scenarios related to technological innovations, policy interventions, climate change and meteorological variability, and social and economic issues. This paper addresses further development of the previously published Ebbsfleet Water Management SDM, co-produced with the Ebbsfleet LAA. This is a new development, built on previous work where the structure of the SDM was largely determined by the stakeholders. While the participatory process facilitated by the LAA was very valuable, the stakeholders do not appear to use the final product (aside from potentially reading the published paper), hence there is scope for further enhancement and increase in impact.

This paper demonstrates that the current SDM and the modelling approach used are open to adjustment, which is illustrated by introducing the link between water tariffs and the level of environmental awareness. The introduced changes are not only helpful in providing a more realistic representation of the selected processes, but also, importantly, aim to highlight to the potential end-user that the SDM is open to adjustment, and incentivise usage by stakeholders and further developments both by the stakeholders and the wider modelling community. Further suggestions regarding other potential modifications provide a number of starting points for future research.

5.1 Plans for increasing impact through training events and incorporation in teaching modules

It is worth reiterating that any model is a simplification of reality, and simulation models are limited by their scope (e.g. the imposed system boundaries) and the available research effort. The main value of the system dynamics participatory modelling process, however, is that all the parties involved enhance their understanding of the processes and interactions within the system considered. It is, therefore, stipulated that any modelling product is not rigidly dogmatic but rather open to adjustment, and that the stakeholders are not baffled by the apparent complexity but incentivised to use the product and to develop it further.

The example of simple structural changes to the Ebbsfleet SDM described in this paper provides an optional modification of the previously published model and aims to show to the potential end-user that both the processes represented in the model and their interlinkages can be easily adjusted. The users have full control over the product and can therefore make any modifications they like, including e.g. additions of new components and/or links between the already existing components. The system boundaries could be easily broadened, and the potential new developments may, for example, include representation of indirect impacts on biodiversity, water and air quality, local economy and human health. The output from the Ebbsfleet SDM may also be used as an input to other models.

To facilitate the use of the Ebbsfleet SDM by stakeholders, a series of online training events is currently being considered. These workshops aim to introduce the new users to the basics of simulation modelling and system dynamics, to help with installation and learning the Vensim software interface, to help troubleshoot any teething problems and provide the initial support in running the existing model, and to demonstrate the potential for structural modifications by guiding the users through some examples. The users would then be encouraged to implement further changes, and it is intended to provide feedback and support for their initial attempts. In addition, we are also currently exploring opportunities for using the Ebbsfleet SDM in university teaching. It is planned to compile a distribution bundle containing tutorials for Vensim software and the teaching materials related to the Ebbsfleet SDM. It is deemed that such training materials centred on an interdisciplinary modelling case study should be of use for undergraduate and master level modules related to a wide range of fundamental and applied disciplines, including e.g. Ecology, Geography, Engineering, Social and Environmental Sciences. Furthermore, discussions are currently underway as regards the development of the Environmental Modelling modules for the RBGE MSc course (Edinburgh, Scotland) and the ecology degree in Karazin University (Kharkov, Ukraine). It is intended that an introduction to the Vensim software and the Ebbsfleet SDM should constitute a major part of these modules, alongside other case studies relevant to water management, water quality and aquatic ecology, including the Lake 2 model J. Solomonsen [22] and the model Rostherne [23]–[25].

WIT Transactions on Ecology and the Environment, Vol 251, © 2021 WIT Press
www.witpress.com, ISSN 1743-3541 (on-line)

5.2 Potential for using alternative software and links with other models

The original model is open-source, compiled in the Vensim software, which is free for non-commercial use. It should be noted, however, that although the current version of the model is compiled in Vensim, there is no reason why interested users could not re-implement the model in other interactive modelling software such as Stella, Madonna or Simile. It would also be fairly straightforward to compile the Ebbsfleet SDM in most programming languages (e.g. Fortran, Visual Basic, C, Matlab) given that the model is open-source and all the equations have been documented previously in rather comprehensive detail [14]. In our opinion, Simulink and Matlab might be particularly convenient for the re-implementation and further development of the Ebbsfleet SDM (however, NB the costs). It should also be noted that the current definitions of the model contain only four differential equations (the rest are algebraic). By approximating those ordinary differential equations by difference equations, the model could be sufficiently simplified for an easy implementation in a spreadsheet, such as e.g. 'Calc' (for Linux), 'Numbers' (for Mac) or 'Microsoft Excel' (for Windows). The latter software in addition to the spreadsheet capabilities also has a built-in version of BASIC (i.e. VBA), and may, therefore, be particularly useful for debugging the spreadsheet model and comparing its output with the version based on differential equations.

The re-implementation and further development of the Ebbsfleet model in a programming language would help to overcome some of the major limitations of SDM. The model could be linked to a GIS, thus overcoming the inability to represent spatial processes. Also, the flexibility of most programming languages would allow easy access to a plethora of scripts shared by their user community. This, as well as an opportunity for writing new functions and sub-programs, would also create opportunities for representation of finer scale dynamics thus allowing another major constraint of SDM to be overcome and creating a more flexible, custom-built output. In particular, we envisage the scope of linking the adapted SDM to Urban Water Metabolism models. Such coupling would be very beneficial and would help to overcome the usual limitations in developing a detailed metabolic model e.g. data availability and uncertainty at high resolution. Furthermore, the system dynamics modelling approach may be particularly valuable for the analysis of indirect interrelations among system constituents thus enhancing flexibility of the water metabolism models.

5.3 Overall relevance

The optimisation of urban water management is key to the sustainable development of modern cities [4], [14]. System dynamics studies are very useful in that respect and have a potential added value of engaging practitioners and the public in further developments of the resulting models. Consequently, all users enhance their understanding of the underlying complexity through 'learning by doing' [26] and 'learning by example' [27], and ultimately, society benefits from the increased knowledge of its members. The research presented here has practical implications in helping to achieve and promote these intangible societal values, and is therefore contributing to the ongoing development of the Blue-Green Cities conceptual framework [1]–[3], [28], [29].

ACKNOWLEDGEMENTS

This research was part of an interdisciplinary project undertaken by the UFR Research Consortium (www.urbanfloodresilience.ac.uk). This work was supported by the UK Engineering and Physical Sciences Research Council (grant numbers EP/P004180/1, EP/P003982/1, EP/P004318/1). Colin Thorne is thanked for valuable discussion and edits.

REFERENCES

[1]   O'Donnell, E. et al., The blue-green path to urban flood resilience. *Blue-Green Systems*, **2**(1), pp. 28–45, 2020.

[2]   Fenner, R. et al., Achieving urban flood resilience in an uncertain future. *Water*, **11**(5), 2019.

[3]   D'Arcy, B. J., Kim, L.-H. & Maniquiz-Redillas, M., *Wealth Creation Without Pollution. Designing for Industry, Ecobusiness Parks and Industrial Estates*, London: IWAP, 2018.

[4]   Pagano, A., Pluchinotta, I., Giordano, R. & Fratino, U., Integrating "hard" and "soft" infrastructural resilience assessment for water distribution systems. *Complexity*, **16**, 2018.

[5]   Pagano, A., Pluchinotta, I., Pengal, P., Cokan, B. & Giordano, R., Engaging stakeholders in the assessment of NBS effectiveness in flood risk reduction: A participatory System Dynamics Model for benefits and co-benefits evaluation. *Science of the Total Environment*, **690**, pp. 543–555, 2019.

[6]   Pluchinotta, I., Pagano, A., Giordano, R. & Tsoukiàs, A., A system dynamics model for supporting decision-makers in irrigation water management. *Journal of Environmental Management*, **223**, pp. 815–824, 2018.

[7]   Behzadian, K. & Kapelan, Z., Modelling metabolism based performance of an urban water system using WaterMet2. *Resources, Conservation and Recycling*, **99**, pp. 84–99, 2015.

[8]   Venkatesh, G., Brattebø, H., Sægrov, S., Behzadian, K. & Kapelan Z., Metabolism-modelling approaches to long-term sustainability assessment of urban water services. *Urban Water Journal*, **14**(1), pp. 11–22, 2017.

[9]   Zhuge, C., Yu, M., Wang, C., Cui, Y. & Liu, Y., An agent-based spatiotemporal integrated approach to simulating in-home water and related energy use behaviour: A test case of Beijing, China. *Science of the Total Environment*, **708**, p. 135086, 2020.

[10] Correia de Araujo, W., Oliveira Esquerre, K.P. & Sahin, O., Building a system dynamics model to support water management: A case study of the semiarid region in the Brazilian Northeast. *Water*, **11**(12), p. 2513, 2019.

[11] Chhipi-Shrestha, G., Hewage, K. & Sadiq, R., Economic and energy efficiency of net-zero water communities: System dynamics analysis. *Journal of Sustainable Water in the Built Environment*, **4**(3), p. 04018006, 2018.

[12] Tidwell, V.C., Passell, H.D. & Conrad, S.H. & Thomas, R.P., System dynamics modeling for community-based water planning: Application to the Middle Rio Grande. *Aquatic Sciences*, **66**(4), pp. 357–372, 2004.

[13] Jeong, S. & Park, J. Evaluating urban water management using a water metabolism framework: A comparative analysis of three regions in Korea. *Resources, Conservation and Recycling*, **155**, p. 104597, 2020.

[14] Pluchinotta, I. et al., A participatory system dynamics model to investigate sustainable urban water management in Ebbsfleet Garden City. *Sustainable Cities and Society*, 67, p. 102709, 2021.

[15] CIRIA, *Blue-Green Infrastructure: Perspectives on Water Quality Benefits*, London: CIRIA, C780b, 2019.

[16] Krivtsov, V., Corliss, J., Bellinger, E. & Sigee, D., Indirect regulation rule for consecutive stages of ecological succession. *Ecological Modelling*, **133**(1–2), pp. 73–81, 2000.

[17] Krivtsov, V., Investigations of indirect relationships in ecology and environmental sciences: A review and the implications for comparative theoretical ecosystem analysis. *Ecological Modelling*, **174**(1–2), pp. 37–54, 2004.

[18] Krivtsov, V., Study of cause-and-effect relationships in the formation of biocenoses: Their use for the control of eutrophication. *Russ J Ecol.* **32**(4), pp. 230–234, 2001.

[19] Krivtsov, V., Indirect effects in ecology. *Encyclopedia of Ecology*, ed. S.E. Jorgensen & B. D. Fath, Newnes, pp. 1948–1958, 2008.

[20] Ahilan, S. et al., Modelling the long-term suspended sedimentological effects on stormwater pond performance in an urban catchment. *Journal of Hydrology*, **571**, pp. 805–818, 2019.

[21] Woods-Ballard, B., Kellagher, R., Martin, R., Jefferies, C., Bray, R. & Shaffer, P., *The SuDS Manual*, London: CIRIA C697, 2007.

[22] Jørgensen, S.E., *Fundamentals of Ecological Modelling*, 2nd ed., Elsevier: Amsterdam, 1994.

[23] Krivtsov, V., Bellinger, E., Sigee, D. & Corliss, J., Interrelations between Si and P biogeochemical cycles: A new approach to the solution of the eutrophication problem. *Hydrological Processes*, **14**(2), pp. 283–295, 2000.

[24] Krivtsov, V., Goldspink, C., Sigee, D.C. & Bellinger, E.G., Expansion of the model 'Rostherne' for fish and zooplankton: Role of top-down effects in modifying the prevailing pattern of ecosystem functioning. *Ecological Modelling*, **138**(1–3), pp. 153–171, 2001.

[25] Krivtsov, V., Sigee, D., Corliss, J. & Bellinger, E., Examination of the phytoplankton of Rostherne Mere using a simulation mathematical model. *Hydrobiologia*, **414**, pp. 71–76, 1999.

[26] Schank, R.C., Berman, T.R. & Macpherson, K.A., Learning by doing. *Instructional-Design Theories and Models: A New Paradigm of Instructional Theory*, ed. C.M. Reigeluth, vol. 2, Lawrence Earlbaum Associates, pp. 161–181, 1999.

[27] Bandura, A., *Social Learning Theory*, General Learning Press: New York, 1971.

[28] CIRIA, *Blue-Green Infrastructure: Perspectives on Planning, Evaluation and Collaboration*, London: CIRIA, C780a, 2019.

[29] Krivtsov, V. et al., Flood resilience, amenity and biodiversity benefits of an historic urban pond. *Philosophical Transactions of the Royal Society A*, **378**(2168), p. 20190389, 2020.

HOW MANAGEMENT EFFORTS OF A LARGE HYDROPOWER FIRM IMPACT SUSTAINABILITY IN THE COLOMBIAN ANDES: A MULTIVARIATE ANALYSIS OF SOCIETY'S PERCEPTIONS

DANIEL CORTÉS-BORDA, JORGE-ANDRÉS POLANCO & MANUELA ESCOBAR-SIERRA
Faculty of Economic and Administrative Sciences, University of Medellin, Colombia

ABSTRACT

When it comes to sustainability, watersheds and hydropower firms must be conceived as a whole. Namely, hydropower dams impact the three dimensions of sustainability of watersheds, while dams' lifetime is lowered by unsustainable practices taking place in the watersheds. Management of hydropower firms aiming at sustainability might ensure the long-term use of dams without compromising ecosystems and society's welfare. We aim to assess the impact of management efforts of a large hydropower firm on the sustainability of the influenced watersheds from the perceptions of society. We build on survey data assessing the social perception of the impacts caused by a large hydropower plants operation; and the firm's management efforts aiming at sustainability. To this end, we perform a stepwise multilinear regression of ad-hoc impact management indices (independent variables) and impact indices (dependent variables). Data comprises more than 600 surveys from community, policymakers and industry, from two watersheds in Colombian Andes. Results revealed a positive correlation between all the impact indices and the management indices concerning environment and economy-society. The remaining management indices showed no (or low) correlation with impacts. Findings suggest that, despite firm's sustainability awareness, society perceives low positive impact in due to (what they consider) firm's few impact management efforts in environmental indices (i.e., erosion and deforestation) and socio-economic indices (i.e., income alternatives in agriculture, fisheries and tourism activities). Other efforts are not perceived as (positive or negative) consequences of the impact. Correlation results provided valuable information, for scholars and practitioners, on the interaction of dams and watersheds. On one hand, the theoretical implications showed how a holistic approach of sustainability is needed to better understand the complexity of this relationship. On the other, the management implications gave insights on how a large hydropower plant can operate in the long term while causing a positive impact on ecosystems and the local society.
Keywords: sustainability, hydropower, stepwise regression, social perception, watershed, social impact, environmental impact.

1 INTRODUCTION

The United Nations Sustainable Development Goals (SDGs) concerning energy aim to achieve global access to energy, reduce the severe health impacts of air pollution, and tackle climate change, which altogether might be achieved by substantially increasing the share of renewable energies. Hydropower, today's primary renewable energy source, plays a crucial role in meeting the SDGs [1] since it helps to mitigate greenhouse gas emissions while reducing reliance on imported fuels, and it is an inexpensive power source that adapts to the different needs and possibilities of the emerging and developed economies [2]. Moreover, storing water in hydroelectric dams provides grid stability [3] and offers other advantages [4]–[7] that are particularly interesting for emerging economies (e.g., drought management, irrigation, water supply, flood control, aquaculture, tourism and other job opportunities).

WIT Transactions on Ecology and the Environment, Vol 251, © 2021 WIT Press
www.witpress.com, ISSN 1743-3541 (on-line)
doi:10.2495/WS210031

Although hydropower has numerous benefits, it has been intensely criticized [8]–[10] due to its social and environmental adverse effects [11]–[14] related to its extractive nature [15]. That is, the construction and operation of dams lead to population relocation, social conflicts, landslides, habitat changes, among other relevant issues that threaten social and biodiversity welfare [15], [16]. It is well known that the social impact of large hydroelectric dams leads to social transformations [17] derived from the inequitable distribution of positive and negative impacts among the stakeholders [18]. Conflicts between hydropower firms and society have raised questions related to the sustainability of dams and the rights of the watershed's inhabitants [19]–[21]. Moreover, emerging economies are particularly vulnerable to climate induced hydrological [22], [23] changes that threaten hydropower projects (i.e., erosion and sedimentation that shorten their lifetime and their production capacity [24]).

Lessons learned from the past have taught that social and environmental issues have restricted the hydropower expansion projects of two of the world's leading hydropower producers, China and Brazil [1]. Hence, the management of hydropower projects require guidance in environmental, social, financial, and technical sustainability criteria, from the joint perspective of industry, civil society, policy makers, and financiers [25]–[27]. Practical solutions to overcome hydropower drawbacks include developing of protective measures [26]; performing comprehensive impact assessments [7], risk evaluations and management plans involving the stakeholders [29], [30]; engaging stakeholders in the decisions related to the watershed [31]–[33]; and a benefit sharing scheme [34], [35]. However, adopting a holistic sustainability approach in the management of large hydroelectric firms [11] (e.g., corporate sustainability [36]) could be a solution to create value while strengthening the social and environmental development of watersheds.

Colombia is an emerging economy with a large hydropower potential. The growing energy demand along with the national government priorities [16] placed Colombia among the top hydropower producers in the world [37], and it is expected to continue expanding [38]. Unfortunately, the intensive deployment of hydropower in Colombia has led to severe accidents [39] and environmental, social, and armed conflicts [15], [16], [18], [40], [41] often derived from deep inequalities [42], the lack of trust of the stakeholders and the weak interventions of the state [43].

The sustainability of the hydropower sector in Colombia has gained increasing awareness and financial support [44]. In recent years, sustainability issues related to hydropower in Colombia has been assessed, i.e., the management of sediments [45], [46], the sediments generation according to the land use [47], the payment for ecosystem services [48], the benefit sharing scheme [35], the governance for sustainability in a context of violence [20], the equity and sustainability of water allocation [49] and the political events that have led to social conflicts related to hydropower [16]. Other authors have modelled different scenarios and proposed policies to aid in the decision-making process [50]. In terms of corporate sustainability, Polanco [43] identified the relationship between the strategy and the political stance of the hydropower companies. Despite the numerous studies related to the sustainability of hydropower in Colombia, it is still necessary to establish a quantitative cause–effect relationship between sustainability and the management actions taken in hydropower firms.

In a recent project intitled "A new measurement system design for monitoring sustainability performance of hydropower" our research team surveyed more than 600 society members (including community, policymakers and industry) from two different watersheds in Colombian Andes to assess the social perception of the impacts caused by the operation of two large hydropower plants and the firm's management actions to tackle such impacts. Going beyond the surveys, we aim to assess the impact of management efforts of a large hydropower firm on the sustainability of the influenced watersheds from the perceptions of society. To this end, we performed a stepwise multilinear correlation of ad-hoc sustainability management indices (independent variables) and impact indices (dependent variables) to establish a correlation that reveals valuable information (for scholars and practitioners) on the effectiveness of the firm's actions. To the best of our knowledge, this is the first study (in the Colombian Andes) establishing a cause–effect quantitative relationship between the impact of hydroelectric dams and the management efforts aiming at sustainability.

The remainder of this document is organized as follows. Section 2 establishes the hypothesis of our research, followed by a detailed description of the nature of data and the description of the statistical methods here used. Section 3 presents the main findings of the stepwise correlations and its implications, while conclusions that might be profitable for scholars and practitioners are drawn in Section 4.

2 METHODS

In this work we aim to establish cause–effect relationships between impact management efforts and sustainability. To do so, here we formally state the hypothesis to be tested through a stepwise regression of ad-hoc sustainability indices. Then, we provide details on the survey data and the construction of impact indices (assessing the social perception of the impacts caused by two large hydropower plants), and the impact management indices (assessing the social perception of the firm's management efforts to tackle such impacts). We examine cause–effect relationships by performing stepwise correlations of the impact and impact management indices (according to their location and social segment).

2.1 Hypothesis

The hypothesis that we aim to probe is as follows:

H1: Colombian Andes society perceives the impact of management efforts to improve the sustainability of the watersheds of a large hydropower firm on the three dimensions of sustainability, and its interactions.

2.2 Data

The data set consists of 694 social perception surveys made in 2018. Data belongs to two different watersheds of the Colombian Andes (R1 and R2, with 377 and 317 surveys, respectively), and three different social segments: Community (Com), Policymakers (Pol), and Industry (Ind) (with 605, 68 and 21 surveys, respectively). The measuring instrument comprises 25 questions measuring the perception in 1–5 Likert scale, where 5 is the most positive perception. A translated to English version of the 25-question survey used as measuring instrument is provided in the Appendix. The 25 questions are divided as follows: questions Q1–Q10 assess the social perception of impact, questions Q11–Q25 assess the social perception of impact management. The resulting 25 questions concerning evaluate two of the three dimensions of sustainability, and the interactions of the three dimensions of sustainability. For the sake of simplicity, we refer to the ensemble of the (aforementioned) two dimensions of sustainability and the three interactions as "categories."

We rely on the categories previously defined by Polanco [20], which are as follows: the environmental category (Env) concerns climate change, biodiversity, and the condition of natural resources, water, and soil; the social category (Soc) refers to building social capital, trust, relationships, and organizational networks; the environment-economy category (Env–Eco) is related to the economy of natural resources (supply and demand), pollution of the environment, and basic sanitation; the economy-society category (Eco–Soc) is associated to activities for subsistence, food security, production organization, and commercialization of farm products; and the environment-society category (Env–Soc) entails training, consciousness, and environmental culture in the territory. Noteworthy, the economic dimension of sustainability was not addressed here since it was studied in deep by Polanco et al. [51].

2.3 Indices construction and preliminary analysis

We start by treating the results of the 25 questions as variables. We validated an eventual stepwise correlation by proving their linear independence via the rank of the matrix of variables (25 questions and 694 observations). Since, linear regressions with a large set of variables may lead to poor model predictions, we attempted to reduce the dimension of the problem without a significant loss of information by grouping variables through a principal component analysis (PCA). The PCA did not lead to linear combinations of variables leading to a reduced model (which fits the results of the linear independence test).

Since variables were not significantly correlated, we grouped variables according to their categories, leading to five impact indices (I1–I5) and five impact management indices (I6–I10), i.e., one impact index and one impact management index per category. Fig. 1 depicts the data structure and how questions are grouped (according to their category) to create impact indices and impact management indices.

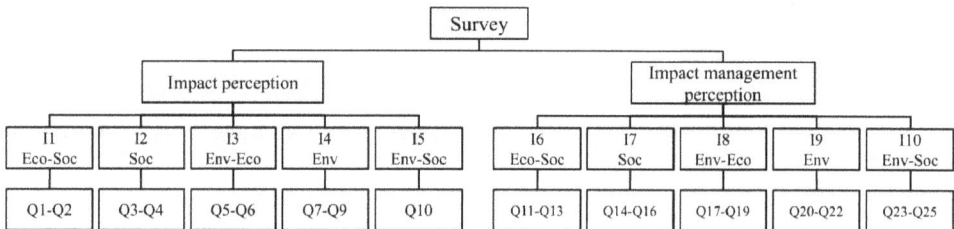

Figure 1: Data structure and indices construction.

Indices are calculated as the arithmetic mean of all the variables belonging to the same category (see Fig. 1). As an example, let us consider index I1 (perception of impact in the socioeconomic category). The index I1 is a 694-element column vector calculated as the average of columns corresponding to questions Q1 and Q2, where questions Q1 and Q2 represent the impact perception of the 694 surveyed individuals, in the Eco–Soc category. We verified that there is no multicollinearity in the new matrix of indices (10 indices and 694 observations) by calculating the rank the matrix.

As a complementary analysis of the stepwise regression, we compared the average level of perception in each index and explored if there are significant differences in the perception level of impact and impact management according to the geographical locations and the social segments via two-way ANOVA.

2.4 Stepwise regression

Stepwise regression is a well-known method that contrasts numerous multivariate linear regression models and selects the one with the best-fit of data [52]. The algorithm starts with an initial model and makes successive steps to improve it by adding and removing terms according to the significance of the independent variables (which is determined by an F-test hypothesis in each linear regression). In each step, the stepwise regression method uses the least squares method to estimate the model coefficients. The stepwise regression is an iterative process that terminates when the model does not improve with further steps.

We used the MATLAB built-in stepwise regression algorithm to validate our hypothesis. In this correlation, the impact perception indices (I1–I5) represent the dependent variables and the impact management perception indices (I6–I10) represent the independent variables.

Thereby, per each impact perception index, the stepwise regression method selects which of the five independent variables make significant contributions to the multilinear model. Since the perception of impacts and impact management might be significantly different according to the geographical location and social segment, we examined one stepwise regression for each impact index, for each region, and for each social segment (including the global set of data, denoted as 'All'). This approach leads to 30 stepwise multilinear models shedding light on the cause–effect relationship between the impact management and the sustainability impacts derived from the intervention of hydropower in the Colombian Andes.

3 RESULTS AND DISCUSSION

In a preliminary analysis we compared the average perception values in the studied groups, which revealed that, as a general trend, the industry has the most favorable perception of impact and impact management, while the community is the most pessimistic social segment. Regarding the geographical location, in most of the impact and impact management indices, Z1 exhibited a better perception than Z2. At first glance, we observed differences in the perception level among groups, however the two-way ANOVA assessment discloses which of the differences are statistically significant.

Results from the two-way ANOVA showed that the perception of Z2-Com, the most pessimistic group, is significantly different from the other groups in all the indices, except for I10 (impact management index related to training, consciousness, and environmental culture in the territory). The ANOVA results also showed that, excluding Z2-com, all the significant differences in perception are related to impact management indices, and not in impact indices. That is, the group Z1-com presents significant differences in indices I6, I7, I8, I9; the group Z1-Pol is significantly different in index I8; and the groups Z1-Pol and Z2-Pol presents differences in I6. Later in this section, we will develop joint analysis of the results of the two-way ANOVA and the stepwise regressions.

Once we validated that there is no multicollinearity in the matrix of indices (10 indices and 694 observations) by calculating the rank of the full data matrix, we performed 30 multilinear stepwise regression models considering different geographical locations and social segments. The alpha value used to contrast the hypothesis was 1% (e.g., variables considered in each regression model have a significance level of, at least, 99%).

The results of the coefficient of determination (R^2) of each stepwise correlation model (following a color scale) are given in Table 1. It is worth mentioning that we aim to identify cause–effect relationships between the impact management indices and the impact indices, and do not aim to predict a precise value of the impact as a function of the impact management. Hence, the relatively low values of the R^2 are not a major concern for the scope of this study.

Table 1: Coefficient of determination (R^2) of each stepwise model.

Social segment	Coefficient of determination (R^2)					
	I1 Eco–Soc	I2 Soc	I3 Env–Eco	I4 Env	I5 Env–Soc	Average
Ind		0.39	0.20		0.47	0.35
Pol		0.09	0.19	0.14	0.12	0.13
Com	0.38	0.38	0.36	0.37	0.28	0.35
Z1	0.27	0.27	0.35	0.26	0.19	0.27
Z2	0.30	0.30	0.26	0.38	0.25	0.30
All	0.37	0.38	0.37	0.38	0.29	0.36
Average	0.33	0.30	0.29	0.31	0.27	

Remarkably, three out of the 30 models are absent. No (or low) correlations entail that the studied social segment does not perceive any (positive or negative) effect of the management efforts to improve sustainability in the sustainability performance of the watershed. Industry and policymakers do not perceive a direct relationship between the Eco–Soc impact and any of the impact management indices. Similarly, the industry does not perceive correlations between the environmental impact and any impact management index. In addition, by comparing the R^2 average values, we observe that the policymakers' segment presents the weakest cause–effect correlations. The fact that industry and policymakers exhibit low and absent correlations could be explained due to the local economic activities are not directly impacted by the operation of the hydroelectric dams.

We present in Table 2 the stepwise regression models (i.e., variable coefficients and intercept). By assessing the frequency of the impact management indices that appeared in correlations, we observed that I6 (Eco–Soc) and I9 (Env) appeared in 60% of the correlation models, followed by I10 (Env–Soc) which is present in 23.3% of the models, and I7 (Soc), and I8 (Env–Eco) which appeared in 20% of the models.

The frequent appearance of indices I6 and I9 (firm's efforts to improve economic activities of local inhabitants, and firm's efforts in environmental protection and conservation, respectively) in the correlation models could be due to such indices are strongly associated to primary sector economic activities (the main economic activities in Z1 and Z2). Therefore, the firm's actions promoting income alternatives in agriculture, fisheries and tourism activities; as well as the actions to prevent soil erosion and deforestation (which directly affects the availability of primary sector raw materials and products) are perceived by the society as a direct cause of sustainability improvement. By examining correlations of the remaining impact management indices, we observe that, in most of the social segments, the firm's actions building social capital (I7) is perceived by the society as a direct cause of trust and organizational networks improvement (I2), and the promotion of environmental culture (I5) (only in the group All). While the efforts made on the economy of natural resources (supply and demand), pollution of the environment, and basic sanitation (I8) are perceived (mainly by Z1) as a cause of improvement in social (I2), environmental–economic (I3), environmental (I4) categories. The effect of training, consciousness, and environmental culture in the territory (I10) over indices I1, I3, I4 and I5 appears to be conflicting since there are positive and negative correlations with the impact indicators.

Table 2: Stepwise regression models results.

Impact indicator	Social segment	Coefficient values					
		I6 Eco–Soc	I7 Soc	I8 Env–Eco	I9 Env	I10 Env–Soc	Intercept
I1 Eco–Soc	Ind						
	Pol						
	Com	0.77			0.23		−0.04
	Z1	0.61			0.33		0.15
	Z2	0.76				0.25	−0.26
	All	0.74			0.21		0.11
I2 Soc	Ind		0.60				1.62
	Pol		0.43				2.12
	Com	0.50	0.30	0.20			0.60
	Z1	0.43		0.33			1.35
	Z2	0.40	0.36				1.12
	All	0.46	0.34				1.16
I3 Env–Eco	Ind				0.74		1.26
	Pol					0.53	1.81
	Com	0.52			0.40		0.78
	Z1	0.24		0.28	0.42		0.75
	Z2	0.42			0.43	−0.21	1.52
	All	0.50			0.39		0.86
I4 Env	Ind						
	Pol				0.52		1.60
	Com	0.38		0.24	0.40	−0.24	0.67
	Z1	0.33		0.30		0.23	0.33
	Z2	0.40			0.49	−0.21	0.97
	All	0.38		0.18	0.40	−0.19	0.68
I5 Env–Soc	Ind	0.93					0.45
	Pol				0.31		3.01
	Com	0.14			0.66		1.27
	Z1				0.39	0.35	1.35
	Z2				0.69		1.40
	All		0.13		0.62		1.38
Average		0.49	0.36	0.25	0.45	0.06	

Next, we examined the magnitude and sign of the variable coefficient values. A fast-track approach to compare the contribution of variables to the models consists of comparing the average of the variable coefficients (excluding zeroes). From this comparison we observed that, I6 and I9 are not only appearing frequently but also their contribution is higher than other variables. Noteworthy, most of the variable coefficients shown in Table 2 are positive, except some of the variable I10 (Env–Soc) (in I3 Z2 and I4 Com, Z2 and All). The positive correlations among cause–effect variables were expected since impact management efforts are intended to improve sustainability. The few negative correlations require more studies to disclose whether an eventual reverse effect of the management actions aiming at sustainability. However, since the magnitude of the negative variable coefficients is low (the average variable of the I10 variable coefficient is close to zero), the negative values could be due to noise, which would imply no correlation of I10 with I1–I5 instead of a reverse effect between actions and sustainability.

Finally, we performed a joint analysis of the ANOVA and stepwise regression results to link the correlation level with the level of positive perception. Noteworthy, social groups with highly positive perception (i.e., policymakers and industry) present (in general) weak correlations between impact and impact management. The fact that strong cause–effect correlations do not appear in groups with a better perception entails that their positive perception cannot be ascribed to the effectiveness of the firm's management efforts. Moreover, the absent correlations in such groups may suggest that the impact management efforts done in the indicators that are not correlated are not even perceived by these social groups. The joint analysis also revealed that the most pessimistic groups (e.g., group Z2-Com) perceives an evident cause-effect relationship between sustainability and impact management (i.e., present high correlation values between impact and impact management). This may suggest that, despite the firm's sustainability awareness and the significant efforts made to improve the sustainable development in the watershed, these groups perceive that the impact management efforts in are low.

4 CONCLUSIONS

After performing a stepwise regression model, we can confirm our hypothesis. In fact, Colombian Andes society perceives the impact of management efforts to improve the sustainability of the watersheds of a large hydropower firm on the three dimensions of sustainability, and its interactions.

Information retrieved from this study could shed light on where to concentrate more impact management efforts, and which efforts might be evaluated (since they are not being perceived). Particularly, impact management efforts perceived as direct cause of improvement of the local sustainability are those made in improving the economy of natural resources, reducing environmental pollution; making basic sanitation; preventing climate change; and protecting biodiversity, natural resources, water, and soil. Whereas management efforts that are less evidently perceived as a cause of sustainability improvement are those made on strengthening social capital, trust, relationships, organizational networks; improving activities for subsistence, food security, production organization, commercialization of farm products; and training, creating consciousness, and environmental culture in the territory.

Despite the hydropower firm's increasing awareness on sustainability, local inhabitants still do not perceive the firm's impact management actions as an attempt to improve the sustainability of the region. Making progress in sustainability would require a shared awareness of sustainability among all the stakeholders.

Correlation results provided valuable information, for scholars and practitioners, on the interaction of dams and watersheds. On one hand, the theoretical implications showed how a holistic approach of sustainability is needed to better understand the complexity of this relationship. On the other, the management implications gave insights on how a large hydropower plant can operate in the long term while causing a positive impact on ecosystems and the local society.

ACKNOWLEDGEMENTS

The authors would like to acknowledge Josep-Maria Mateo Sanz, from the Rovira i Virgili University, Tarragona, Spain, who provided support in the preliminary statistical analysis. The authors would like to thank to the Colombian Ministry of Science, Technology and Innovation (Minciencias) for the funding 80740-663-2020.

APPENDIX

Table A1: Description of question subjects and indices. (Note: textual questions are not provided in Table A1 since the same subjects were differently written according to the social group).

Type of indices	Indices	Question subject
Impact perception indices	I1: Eco–Soc Activities for subsistence, food security, production organization, and commercialization of farm products	Q1: Income increase
		Q2: New income sources
	I2: Soc Building social capital, trust, relationships, and organizational networks	Q3: Participation in decisions
		Q4: Trust in neighbours
	I3: Env–Eco The economy of natural resources (supply and demand), pollution of the environment, and basic sanitation	Q5: Forest and soil protection
		Q6: Water quality and biodiversity availability
	I4: Env Climate change, biodiversity, and the condition of natural resources, water, and soil	Q7: Amount of fish in water bodies
		Q8: Landslidings reduction
		Q9: Overflows reductions
	I5: Env–Soc Training, consciousness, and environmental culture in the territory	Q10: Nature caring awareness
Impact management perception indices	I6: Eco–Soc Activities for subsistence, food security, production organization, and commercialization of farm products	Q11: Improvement of economic activities
		Q12: Tool promoting economic activities
		Q13: Players promoting economic activities
	I7: Soc Building social capital, trust, relationships, and organizational networks	Q14: Participation in local decisions
		Q15: Trust in players
		Q16: Tools motivating trust in actors
	I8: Env–Eco The economy of natural resources (supply and demand), pollution of the environment, and basic sanitation	Q17: Nature use
		Q18: Tools leading to good practices in nature use
		Q19: Players involved in the good use of nature
	I9: Env Climate change, biodiversity, and the condition of natural resources, water, and soil	Q20: Nature conservation
		Q21: Tools promoting nature conservation
		Q22: Players promoting nature conservation
	I10: Env–Soc Training, consciousness, and environmental culture in the territory	Q23: Nature awareness
		Q24: Tools promoting nature awareness
		Q25: Players promoting nature awareness

REFERENCES

[1]   IEA, *Hydropower*, IEA: Paris, 2020. https://www.iea.org/reports/hydropower.

[2]   Bartle, A., Hydropower potential and development activities. *Energy Policy*, **30**(14), pp. 1231–1239, 2002. DOI: 10.1016/S0301-4215(02)00084-8.

[3]   Rehman, S., Al-Hadhrami, L.M. & Alam, M.M., Pumped hydro energy storage system: A technological review. *Renew. Sustain. Energy Rev.*, **44**, pp. 586–598, 2015. DOI: 10.1016/j.rser.2014.12.040.

[4]   Meng, Y., Liu, J., Wang, Z., Mao, G., Wang, K. & Yang, H., Undermined co-benefits of hydropower and irrigation under climate change. *Resour. Conserv. Recycl.*, **167**, 2021. DOI: 10.1016/j.resconrec.2020.105375.

[5]   Meng, Y. et al., Hydropower production benefits more from 1.5°C than 2°C climate scenario. *Water Resour. Res.*, **56**(5), 2020. DOI: 10.1029/2019WR025519.

[6]   Zhou, Y. et al., A comprehensive view of global potential for hydro-generated electricity. *Energy Environ. Sci.*, **8**(9), pp. 2622–2633, 2015. DOI: 10.1039/C5EE00888C.

[7]   Shaktawat, A. & Vadhera, S., Risk management of hydropower projects for sustainable development: A review. *Environ. Dev. Sustain.*, **23**(1), pp. 45–76, 2021. DOI: 10.1007/s10668-020-00607-2.

[8]   Ngor, P.B., Lek, S., McCann, K.S. &Hogan, Z.S., Dams threaten world's largest inland fishery. *Nature*, **563**(7730), p. 184, 2018. DOI: 10.1038/d41586-018-07304-1.

[9]   Gomby, G., Sand in demand: Trapped behind dams. *Science*, **358**(6360), p. 180, 2017. DOI: 10.1126/science.aap9964.

[10] Poff, N.L. & Olden, J.D., Can dams be designed for sustainability? *Science*, **358**(6368), pp. 1252–1253, 2017. DOI: 10.1126/science.aaq1422.

[11] Moran, E.F., Lopez, M.C., Moore, N., Müller, N. & Hyndman, D.W., Sustainable hydropower in the 21st century. *Proc. Natl. Acad. Sci. U.S.A.*, **115**(47), pp. 11891–11898, 2018. DOI: 10.1073/pnas.1809426115.

[12] Yoshida, Y. et al., Impacts of mainstream hydropower dams on fisheries and agriculture in lower Mekong basin. *Sustainability (Switzerland)*, **12**(6), p. 2408, 2020. DOI: 10.3390/su12062408.

[13] Zarfl, C., Lumsdon, A.E., Berlekamp, J., Tydecks, L. & Tockner, K., A global boom in hydropower dam construction. *Aquat. Sci.*, **77**(1), pp. 161–170, 2015. DOI: 10.1007/s00027-014-0377-0.

[14] Winemiller, K.O. et al., Balancing hydropower and biodiversity in the Amazon, Congo, and Mekong. *Science*, **351**(6269), pp. 128–129, 2016. DOI: 10.1126/science.aac7082.

[15] Pérez-Rincón, M., Vargas-Morales, J. & Martinez-Alier, J., Mapping and analyzing ecological distribution conflicts in Andean countries. *Ecol. Econ.*, **157**, pp. 80–91, 2019. DOI: 10.1016/j.ecolecon.2018.11.004.

[16] Martínez, V. & Castillo, O.L., The political ecology of hydropower: Social justice and conflict in Colombian hydroelectricity development. *Energy Res. Soc. Sci.*, **22**, pp. 69–78, 2016. DOI: 10.1016/j.erss.2016.08.023.

[17] Del Bene, D., Scheidel, A. & Temper, L., More dams, more violence? A global analysis on resistances and repression around conflictive dams through co-produced knowledge. *Sustain. Sci.*, **13**(3), pp. 617–633, 2018. DOI: 10.1007/s11625-018-0558-1.

[18] Duarte-Abadía, B., Boelens, R. & Roa-Avendaño, T., Hydropower, encroachment and the re-patterning of hydrosocial territory: The case of hidrosogamoso in Colombia. *Hum. Organ.*, **74**(3), pp. 243–254, 2015. DOI: 10.17730/0018-7259-74.3.243.

[19] Finer, M. & Jenkins, C.N., Proliferation of hydroelectric dams in the Andean Amazon and implications for Andes-Amazon connectivity. *PLoS One*, **7**(4), p. e35126, 2012. DOI: 10.1371/journal.pone.0035126.

[20] Polanco, J.A., Exploring governance for sustainability in contexts of violence: The case of the hydropower industry in Colombia. *Energy Sustain. Soc.*, **8**(1), pp. 1–15, 2018. DOI: 10.1186/s13705-018-0181-0.

[21] Hidalgo-Bastidas, J. & Boelens, R., Hydraulic order and the politics of the governed: The baba dam in coastal Ecuador. *Water*, **11**(3), p. 409, 2019. DOI: 10.3390/w11030409.

[22] Caceres, A.L., Jaramillo, P., Matthews, H.S., Samaras, C. & Nijssen, B., Hydropower under climate uncertainty: Characterizing the usable capacity of Brazilian, Colombian and Peruvian power plants under climate scenarios. *Energy Sustain. Dev.*, **61**, pp. 217–229, 2021. DOI: 10.1016/j.esd.2021.02.006.

[23] Henao, F., Viteri, J.P., Rodríguez, Y., Gómez, J. & Dyner, I., Annual and interannual complementarities of renewable energy sources in Colombia. *Renew. Sustain. Energy Rev.*, **134**, 2020. DOI: 10.1016/j.rser.2020.110318.

[24] Ray, P.A. et al., Multidimensional stress test for hydropower investments facing climate, geophysical and financial uncertainty. *Glob. Environ. Chang.*, **48**, pp. 168–181, 2018. DOI: 10.1016/j.gloenvcha.2017.11.013.

[25] Song, C., O'Malley, A., Zydlewski, J. & Mo, W., Balancing fish-energy-cost tradeoffs through strategic basin-wide dam management. *Resour. Conserv. Recycl.*, **161**, 2020. DOI: 10.1016/j.resconrec.2020.104990.

[26] Siri, R., Mondal, S.R. & Das, S., Hydropower: A renewable energy resource for sustainability in terms of climate change and environmental protection. *Alternative Energy Resources. The Handbook of Environmental Chemistry*, vol. 99, eds. P. Pathak & R.R. Srivastava, Springer: Cham, pp. 93–113, 2020. DOI: 10.1007/698_2020_635.

[27] Liden, R., Specialist, H. & Lyon, K., *The Hydropower Sustainability Assessment Protocol for Use by World Bank Clients: Lessons Learned and Recommendations*, World Bank: Washington, DC, Jun. 2014.

[28] Chen, S., Chen, B. & Fath, B.D., Assessing the cumulative environmental impact of hydropower construction on river systems based on energy network model. *Renew. Sustain. Energy Rev.*, **42**, pp. 78–92, 2015. DOI: 10.1016/j.rser.2014.10.017.

[29] UNEP Dams and Development Project, *Dams and Development: Relevant Practices for Improved Decision-making: A Compendium of Relevant Practices for Improved Decision-making on Dams and their Alternatives*, UNEP-DDP Secretariat: Nairobi, 2007.

[30] Nautiyal, H. & Goel, V., Sustainability assessment of hydropower projects. *J. Clean. Prod.*, **265**, 2020. DOI: 10.1016/j.jclepro.2020.121661.

[31] Choudhury, N.B. & Dey Choudhury, S.R., Implications for planning of hydroelectric projects in Northeast India: An analysis of the impacts of the Tipaimukh project. *GeoJournal*, pp. 1–21, 2020. DOI: 10.1007/s10708-020-10158-8.

[32] Huđek, H., Žganec, K. & Pusch, M.T., A review of hydropower dams in Southeast Europe – Distribution, trends and availability of monitoring data using the example of a multinational Danube catchment subarea. *Renew. Sustain. Energy Rev.*, **117** p. 109434, 2020. DOI: 10.1016/j.rser.2019.109434.

[33] Assessment Protocol (HSAP), Hydropower sustainability tools. https://www.hydrosustainability.org/assessment-protocol.

[34] Wang, C., *A Guide for Local Benefit Sharing in Hydropower Projects*, World Bank: Washington, DC, 2012.

[35] Jiménez-Inchima, I., Polanco, J.A. & Escobar-Sierra, M., Good living of communities and sustainability of the hydropower business: Mapping an operational framework for benefit sharing. *Energy Sustain. Soc.*, **11**(1), p. 9, 2021. DOI: 10.1186/s13705-021-00284-7.

[36] Polanco, J.A. & Ramírez Atehortúa, F.H., *Evaluación de la Sostenibilidad en Empresas de Energía: Una Investigación Aplicada a Centrales de Generación Hidroeléctrica*, Universidad de Medellín, 2017.

[37] Hydropower status report. https://www.hydropower.org/publications/2016-hydropower-status-report.

[38] Quiceno, G. et al., Scenario analysis for strategy design: A case study of the Colombian electricity industry. *Energy Strateg. Rev.*, **23**, pp. 57–68, 2019. DOI: 10.1016/j.esr.2018.12.009.

[39] Sierra, M.C., Ramos, J., Jurado, D. & Herrera, J.D., Practical solutions to geotechnical problems related to Ituango hydropower tunnels, Colombia. *Tunnels and Underground Cities: Engineering and Innovation meet Archaeology, Architecture and Art – Proceedings of the WTC 2019 ITA-AITES World Tunnel Congress*, pp. 1144–1152, 2019.

[40] Pérez-Rincón, M., Vargas-Morales, J. & Crespo-Marín, Z., Trends in social metabolism and environmental conflicts in four Andean countries from 1970 to 2013. *Sustain. Sci.*, **13**(3), pp. 635–648, 2018. DOI: 10.1007/s11625-017-0510-9.

[41] Cernea, M.M., Hydropower dams and social impacts: A sociological perspective (English). Environment Department Working Papers; no. 44, Social Assessment Series*Social Development Papers; no. SDP 16, World Bank Group: Washington, D.C., 1997.

[42] Lindgreen, A., Córdoba, J.-R., Maon, F. & Mendoza, J.M., Corporate social responsibility in Colombia: Making sense of social strategies. *J. Bus. Ethics*, **91**(2), pp. 229–242, 2010. DOI: 10.1007/s10551-010-0616-9.

[43] Polanco, J.A., The group EPM social responsibility: A new political stance against the territory. *Cuad. Adm.*, **27**(49), pp. 65–86, 2014. DOI: 10.11144/Javeriana.cao27-49.

[44] Duque Grisales, E.A., The clean development mechanism as a means to assess the Kyoto protocol in Colombia. *Int. J. Renew. Energy Res. IJRER*, **7**(3), 2017.

[45] Del Río, D.A., Moffett, H., Nieto-Londoño, C., Vásquez, R.E. & Escudero-Atehortúa, A., Chivor's Life Extension Project (CLEP): From sediment management to development of a new intake system. *Water*, **12**(10), p. 2743, 2020. DOI: 10.3390/w12102743.

[46] Del Río, D.A., Moffett, H., Nieto-Londoño, C., Vásquez, R.E. & Escudero-Atehortúa, A., Extending life expectancy of La Esmeralda reservoir: A bet to support Colombia's future energy demand. *American Society of Mechanical Engineers, Power Division (Publication) POWER*, 2020. DOI: 10.1115/POWER2020-16918.

[47] Botero, B.A., Parra, J.C. & Otálvaro, M., Effect of population dynamics and land use on the contribution of sediments to reservoirs for hydropower generation. *WIT Trans. Ecol. Environ.*, **239**, pp. 35–46, 2019. DOI: 10.2495/WS190041.

[48] Rodríguez-de-Francisco, J.C., Duarte-Abadía, B. & Boelens, R., Payment for ecosystem services and the water-energy-food nexus: Securing resource flows for the affluent? *Water*, **11**(6), p. 1143, 2019. DOI: 10.3390/w11061143.

[49] Roa-García, M.C. & Brown, S., Assessing equity and sustainability of water allocation in Colombia. *Local Environ.*, **22**(9), pp. 1080–1104, 2017. DOI: 10.1080/13549839.2015.1070816.

[50] Sierra, R.G. & Sarmiento, Á.Z., New advances in decision making theory under uncertainty and its application in mega projects of hydropower. *World Trans. Eng. Tech. Edu.*, **14**(2), 2016.

[51] Polanco, J.A., Ramírez-Atehortúa, F.H., Montes-Gómez, L.F., Botero-Hernández, B.A. & Barco, M.O., Effect of sediment management decision on a hydropower plant value. *Dyna. Rev. Fac. Nac. Minas*, **87**(213), pp. 232–240, 2020. DOI: 10.15446/dyna.v87n213.81832.

[52] Lütjohann, H., The stepwise regression algorithm seen from the statistician's point of view. *Metr. Int. J. Theor. Appl. Stat.*, **15**(1), pp. 110–125, 1970. DOI: 10.1007/BF02613564.

QUALITY OF GREYWATER IN OMAN AND ITS TREATMENT USING A SUSTAINABLE SYSTEM

MOHAMMED F. M. ABUSHAMMALA, WAJEEHA A. QAZI
& MOHAMMED FAHAD ABDUL LATIF
Department of Civil Engineering, Middle East College, Sultanate of Oman

ABSTRACT
Oman faces the problem of water scarcity owing to the lack of rainfall which results in lower annual replenishment rates in contrast to the consumption rates, and hence requires effective water management. Ablution greywater (AG) is a potential greywater source in Oman as it is less contaminated and discharged in huge amount from mosques. Therefore, this study assessed the quality of AG in Oman to design a sustainable multimedia (sand and activated carbon) filter for its treatment. The coconut shell activated carbon was used in the filter as it provides an inexpensive option due to the huge availability of coconut shell waste in Dhofar region in Oman. All water quality parameters of AG were acceptable as per the Omani standards, except for TSS and turbidity. When compared with past studies in Oman, it was found that the characteristics of AG significantly change with time and therefore should be continuously monitored for effective treatment. The designed multimedia filter adequately removed TSS, turbidity, K, BOD_5, Na, and Cl by up to 96, 96, 61, 50, 24, and 23%, respectively. Initially the concentration of NO_3, SO_4^{-2} and Mg significantly increased, however, later their concentrations started reducing with time. The treatment efficiency fluctuated with time concerning TDS and Ca. The quality of treated AG was not only in compliance with the Omani standards but also WHO guidelines to reuse wastewater for irrigation and toilet flushing. Moreover, by treating 63 m^3 of water monthly, the filter can provide a yearly financial benefit of 519.14 OMR (approximately 1348.48 USD).
Keywords: wastewater, ablution greywater, activated carbon, wastewater treatment.

1 INTRODUCTION
In arid and semi-arid region most of the countries face the problems of water scarcity, prolonged drought and increased demand of water. Being located in this region, Oman also suffers from water shortage. The primary source of water in Oman is groundwater and due to low and highly variable rainfall it faces the tremendous deficit between replenishment and demand. The estimates show that the net annual natural replenishment of groundwater is around 1260 million cubic meters (MCM), whereas the total water demand in Oman is about 1650 MCM. This shows a deficit of 390 MCM which the country accommodates by using groundwater reserves, consequently reducing the water tables which leads to seawater intrusions and salinization in Al Batinah coast, on which 60% of Oman's agriculture rely. The fundamental reason behind the overexploitation of groundwater resources is the perception to meet all water demands by freshwater, be it for drinking or gardening. This perception of water management needs to be changed and country should start focusing on providing sufficient amount of water with acceptable quality. In other words, Oman needs to identify alternative sources of water so as to reduce pressure on freshwater reserves [1].

Greywater originating from bath, laundry and kitchen makes up major part of domestic wastewater, and have low concentration of nutrients, organic material and pathogens [2], [3]. Another type of greywater is the water generated when worshippers perform ablution activity (washing certain body parts) in mosques before praying [3]. Previous studies show that the greywater produced from ablution ritual has even less nutrients, pathogens and organic content as compared to domestic greywater originating from bathing, washing and cooking

WIT Transactions on Ecology and the Environment, Vol 251, © 2021 WIT Press
www.witpress.com, ISSN 1743-3541 (on-line)
doi:10.2495/WS210041

[3]–[5]. According to Radin Mohamed et al. [5], Al-Wabel [6] and Abu-Rizaiza [7], the ablution greywater (AG) has neutral pH (6.92–7.10) with wide variation in BOD (20–40 mg/l), COD (50–70 mg/l), turbidity (10–30 NTU), TSS (5–146 mg/l) and *E. coli* (100–1000 CFU/100 ml). The quality of AG clearly indicates that only a simple treatment is required to recycle it for landscaping, car washing, toilet flushing, irrigation, etc. [3], [4], [8].

Like other Middle Eastern countries, Oman also produces huge amount of greywater from ablution activity. According to Prathapar et al. [9], 1000–1500 L/day of AG is produced from a medium sized mosque in Middle Eastern countries. It is estimated that if 13,000 mosques present in Oman produce AG at the same rate, then daily around 19,500 m^3 of AG is produced which is directly discharged into drainage channels without treatment. Therefore, separating AG at its source and channelling it through basic treatment to recycle for non-potable water application could result in huge amount of water and cost savings [4], [5]. Moreover, large volume of wastewater can be reduced which in turn decrease the cost of sewage treatment plant [10]. However, greywater treatment system also requires extra capital investment, operational and maintenance expenses, thus an effective treatment system based on the quality and quantity of greywater must be designed and adopted which is also financially feasible [9].

Therefore, this study attempts to design, construct and operate a sustainable low-cost multimedia filter for treating AG in Oman, and to investigate its efficiency. This study promotes the use of local and sustainable materials by using activated carbon produced from coconut shell waste, present in huge amounts in the Dhofar region in Oman, which not only saves money but also becomes a way of utilizing waste. The study also determines the change in the characteristics of AG in Oman with time by comparing with previous studies on quality of AG in Oman. Furthermore, the study discusses the capital costs involved in setting up the designed filter, operational and maintenance costs of filter, and potential water and cost savings by recycling AG from the designed system in Middle East College (MEC) mosque.

2 METHODOLOGY

The study was conducted at the mosque located in the campus of MEC, where the AG is discharged directly through drainage pipes into the sewage system. In order to recycle the AG, a treatment system was designed which involved a multimedia filter consisting of silica sand and activated carbon (extracted from coconut shell), as illustrated in Fig. 1. Silica sand and activated carbon were purchased from Amar Sayeed Abdullah Trading LLC and Stability Line Trading LLC, respectively. The multimedia filter was used to remove pollutants and improve the quality of AG so as to reuse it for toilet flushing and irrigation purposes. A cylindrical shaped filter made of fibre reinforced plastic (FRP) was adopted in this study with the height of 1.37 m and diameter of 0.25 m. In the experimental setup, a storage tank filled with AG was connected with the filter. A pump connected to the effluent tube of storage tank pushed AG to the filtration tank to perform filtration. Filtration was carried out at a flow rate of 0.125 m^3/h for 7 days because the daily average flow was 2.2 m^3.

Raw AG was collected from the areas assigned to perform ablution ritual and was immediately transferred to the storage tank connected to the filter for treatment. Samples of AG before and after treatment were collected in a high density polyethylene bottle for chemical analysis and in a sterile sample bottle for microbiological testing. During the filtration period, five before and after treatment samples were collected, one each day, to investigate the quality of AG in terms of pH, BOD$_5$, TSS, TDS, turbidity, NO$_3$, Cl, Na, Mg, K, Ca, SO$_4$$^{-2}$, and *E. coli*.

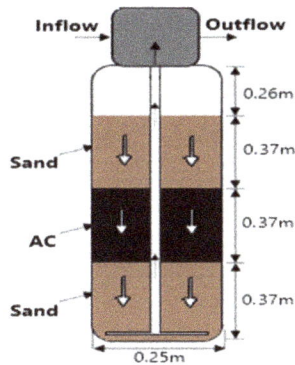

Figure 1: Multimedia filter.

3 RESULTS AND DISCUSSIONS

3.1 Quality of ablution greywater and variation with time

The results shows that pH, BOD_5, TDS, NO_3, Cl, Na, Mg, K, Ca, SO_4^{-2}, and *E. coli* of the generated AG are under the permissible limits set by Omani Standards (Table 1), and can be used for irrigation and toilet flushing. But the concentration of TSS exceeded the allowable limits provided by standard A^1 and A^2. The Omani standard for reuse of wastewater set by MD 145/1993 does not define the permissible limit for turbidity. Similarly, many countries do not have a standard for turbidity and define limit for TSS only. However, it is very important to define maximum allowable limit for turbidity as it can have a negative influence on the performance of irrigation facility, and can reduce soil's hydraulic conductivity which leads to the contamination of soil surface through surface flow. Therefore, a maximum allowable limit of 2 NTU (average) or 5 NTU set by US EPA and Saudi Standards respectively, can be adopted for reuse of treated greywater for irrigation purposes [11]. Consequently, as per the standards, it can be concluded that the quality of AG in terms of TSS and turbidity is not acceptable for reuse and require treatment before reuse.

On comparing the characteristics of AG with the quality of AG reported by Prathapar et al. [9], [12] at Sultan Qaboos University and Al Hail South mosque located in Al Khoudh and Al Hail South areas of Muscat city respectively (Fig. 2), a significant increase is observed in the concentration of TDS, Cl and Na from 2004/2005 to 2019. The highest concentration of TDS was reported in 2004 with 132 mg/l which considerably rose to 536.8 mg/l in 2019. Similarly, an increase of 199.5 mg/l and 126.5 mg/l can be seen in the concentrations of Cl and Na from 2005 to 2019, respectively. The source of freshwater for AG is groundwater, however, the increase in the concentrations of Cl and Na over the years is due to the sea water intrusion problems [13]. However, a significant decline from 58–61 mg/l to 1.35 mg/l is observed in the concentration of biological oxygen demand (BOD_5). Turbidity also reduced from 12.6–51 NTU in 2004/2005 to 5.06 NTU in 2019. Moreover, *E. coli* bacteria was totally absent in the current samples when a concentration ranging from 110 MPN/100 ml to greater than 200 MPN/100 ml was found in the AG in the years 2004 and 2005. These results depicts that the quality of AG significantly changed with time, hence supporting the fact presented by Prathapar et al. [14] on the variation of greywater's quality with time. Therefore, it is considered important to monitor the quality of AG and update the treatment systems with time.

WIT Transactions on Ecology and the Environment, Vol 251, © 2021 WIT Press
www.witpress.com, ISSN 1743-3541 (on-line)

Table 1: Comparison of AG quality from the current study to the previous studies.

Year		2004		2004		2005		2019		Standard A[1a]	Standard A[2a]
Parameter	Unit	SQU		Al Hail South		Al Hail South		MEC			
		Mean	S.D.	Mean	S.D.	Mean	S.D.	Mean	S.D.		
pH		7.10	0.24	7	1.39	7.20	0.16	7.39	0.08	6–9	6–9
BOD$_5$	mg/l	61	22	58	38	–	–	1.35	0.27	15	20
TSS	mg/l	25	22	34	15	9.39	8.20	35.5	16	15	30
TDS	mg/l	120	19	132	28	121	6.86	536.8	51.7	1500	2000
Turbidity	NTU	34	17	51	36	12.6	4.20	5.06	2.24	NA	NA
Cl	mg/l	–	–	–	–	11.46	3.14	211	20.4	650	650
Ca	mg/l	–	–	–	–	19.13	1.50	23.38	10.5	NA	NA
Mg	mg/l	–	–	–	–	0.39	0.05	2.8	2.21	150	150
Na	mg/l	–	–	–	–	9.54	1.80	136	14.9	200	300
K	mg/l	–	–	–	–	4.76	0.39	15.9	4.6	NA	NA
SO$_4^{-2}$	mg/l	–	–	–	–	10.79	8.27	6	1.86	400	400
NO$_3$	mg/l	–	–	–	–	–	–	1.24	0.42	50	50
E.coli	MPN/ 100 ml	110	214	15	40	>200	–	Absent	–	NA	NA

[a]Omani wastewater reuse guidelines as per MD 145/1993 [15].

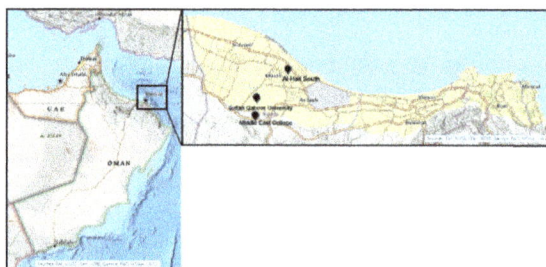

Figure 2: Map of Muscat, Oman with location of mosques.

Furthermore, in 2005, when Prathapar et al. [9] determined the characteristics of AG from Al Hail South mosque in Oman (Fig. 2), he found the quality of AG in terms of pH, COD, EC, TSS and TDS acceptable for reuse; however, the concentration of BOD$_5$, Coliform and E. coli exceeded the permissible limits. In current study, the E. coli was totally absent in the AG, BOD$_5$ was way below the acceptable limit and TSS exceeded the permissible limits, hence showing a totally opposite scenario as compared to the quality of AG back in 2005 in Oman. This also implies that the considerations for designing a treatment system changes with time as the quality of AG varies significantly.

3.2 Quality of treated ablution greywater

Results on the water quality of both untreated and treated AG are given in Table 2. The concentration of turbidity and TSS in AG was higher than the maximum permissible limits, however after treatment the average value of TSS and turbidity reduced to 2.86 mg/l and 1.48 NTU respectively, which are acceptable for irrigation and toilet flushing purposes. Other parameters which were acceptable even before treatment such as, BOD_5, TDS, Cl, Na and K also reduced after treatment. The *E. coli* bacteria was absent in AG before and after the treatment.

Table 2: Comparison of AG before and after treatment.

Parameter	Unit	Untreated		Treated		Standard A[1a]	Standard A[2a]	Acceptable
		Mean	S.D.	Mean	S.D.			
pH		7.39	0.08	9.1	0.314	6–9	6–9	Yes[b]
BOD_5	mg/l	1.35	0.27	0.74	0.134	15	20	Yes
TSS	mg/l	35.5	16	2.86	1.29	15	30	Yes
TDS	mg/l	536.8	51.7	520	20.9	1500	2000	Yes
Cl	mg/l	211	20.4	170.6	14.6	650	650	Yes
NO_3	mg/l	1.24	0.42	2.84	0.93	50	50	Yes
Mg	mg/l	2.8	2.21	19.88	17.32	150	150	Yes
Na	mg/l	136	14.9	108.8	10.26	200	300	Yes
SO_4^{-2}	mg/l	6	1.86	33.56	23.36	400	400	Yes
Ca	mg/l	23.38	10.5	28.32	8.52	NA	NA	
K	mg/l	15.9	4.6	9.46	4.38	NA	NA	
Turbidity	NTU	5.06	2.24	1.48	0.85	NA	NA	
E.coli	CFU/100 ml	Absent	–	Absent	–	NA	NA	

[a]Omani wastewater reuse guidelines as per MD 145/1993 [15].
[b]pH adjustment is necessary.

After treatment the concentration of Mg, Ca, NO_3 and SO_4^{-2} increased, and the possible source for these increases is the use of sand in filter [9]. Despite increasing, the concentration of these chemical contaminants remained well below the maximum permissible levels. Furthermore, pH of the AG also increased after treatment and slightly surpassed the Omani standard threshold for irrigation. The possible reason for the increase in the pH could be the use of coconut shell activated carbon, as the initial effluent produced from most activated carbon products is greater than 7. The initial spike in pH is believed to be due to the formation of positively charged sites on carbon during the activation process. These sites exchange anions with the influent water and raise the pH [16]. This can be seen as the concentration of Cl (anion) also reduced after treatment. Fosso-Kankeu et al. [17] and Mallongi et al. [18] also reported increase in pH level of the water after treatment by activated carbon media. However, this problem can be solved by controlled oxidation of activated carbon surface which stabilize the effluent pH by preventing ion exchange phenomenon [16]. Even though the average pH of treated AG just exceeded the permissible limit by 0.1, it is necessary to adjust the pH before reusing it for irrigation and toilet flushing. Overall, the treated AG was found to be in compliance with the allowable limits of Omani wastewater reuse standards for irrigation and toilet flushing. Moreover, as per WHO guidelines, treated AG quality in this study is suitable for irrigation of fruit trees, fodder crops, and ornamentals.

3.3 Efficiency of treatment unit

The concentration of chemical contaminants in untreated and treated AG, and the removal efficiency of the treatment system are presented in Figs 3 and 4. The multimedia filter adequately removed TSS, turbidity, K, BOD_5, Na, and Cl by up to 96, 96, 61, 50, 24, and

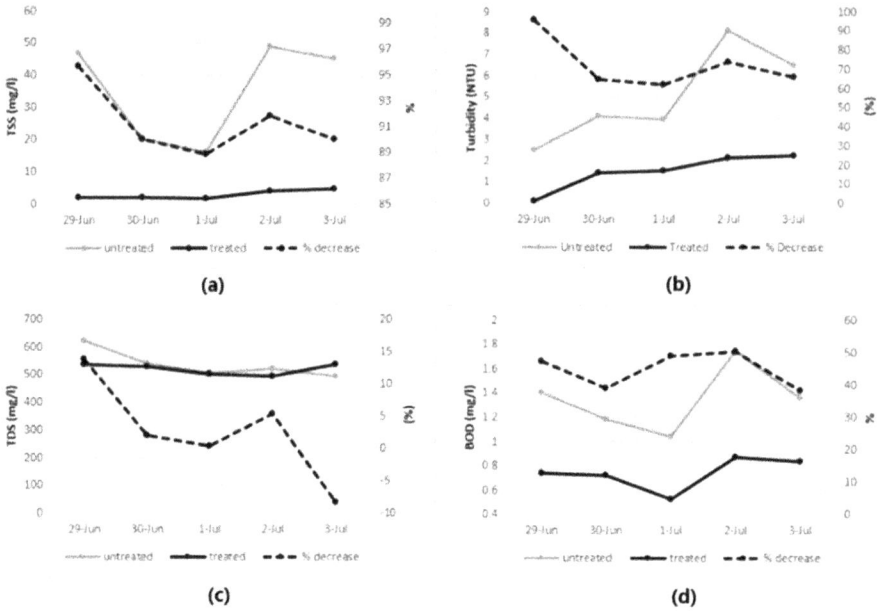

Figure 3: The concentration and removal efficiencies of (a) TSS; (b) Turbidity; (c) TDS; and (d) BOD_5 in untreated and treated AG.

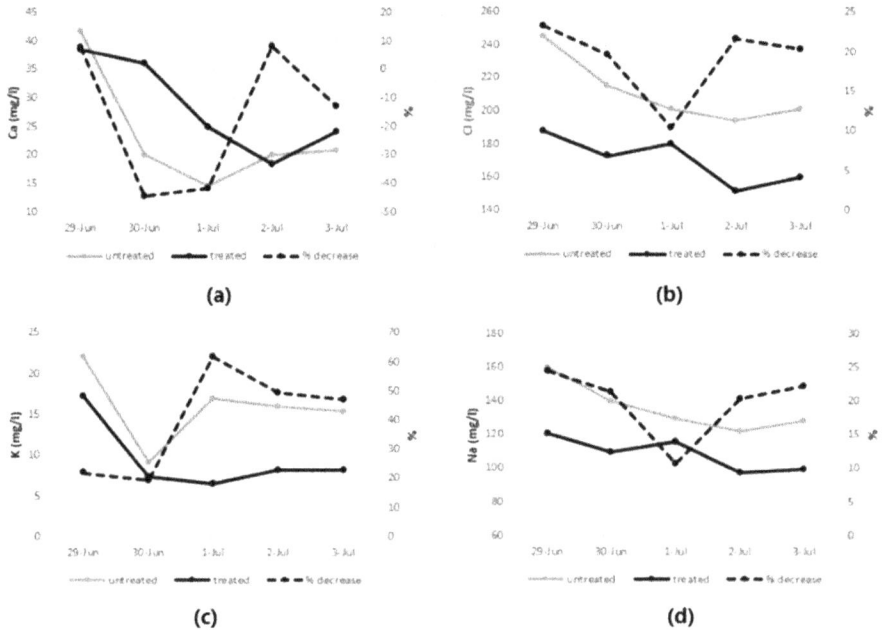

Figure 4: The concentration and removal efficiencies (a) Cl; (b) Ca; (c) K; and (d) Na in untreated and treated AG.

23%, respectively (Figs 3(a), (b), and (d), and 4(b)–(d)). The removal efficiency of these parameters varied slightly with time except for Cl, Na, and K. The removal efficiency of Cl and Na dropped to almost 10% on 1 July; on the other hand removal efficiency for K significantly increased from 19% on 30 June to 61% on 1 July and then slightly varied with time.

The filter achieved the removal efficiency of up to 14% for TDS, which then slightly fluctuated until 2 July before the concentration increased on 3 July by up to 8% (Fig. 3(c)). Similarly, the change in the concentration of Ca in treated water fluctuated considerably with time; the filter removed the concentration up to 8% at one point and also increased the concentration of Ca by up to 44% (Fig. 4(a)).

The concentration of SO_4^{-2} and Mg significantly increased by up to 92% and 99%, respectively. The percent increase in the concentration of SO_4^{-2} and Mg started reducing with time, however on 2 July the concentration of SO_4^{-2} considerably increased again (Fig. 5(b) and (c)). The concentration of NO_3 increased after treatment, but the percent increase in the concentration also started reducing with time (Fig. 5(d)). Prathapar et al. [9] and Al-Zu'bi et al. [10] also reported increase in the concentration of Na, Mg, Ca, carbonates, sulphates, chlorides and nitrate after treating AG. The possible reason for the increase in the concentration of Mg, SO_4^{-2} and NO_3 could be the ion exchange phenomenon that occur in activated carbon, which resulted in the reduction of high quantities of Cl and increase in SO_4^{-2} and NO_3, and decrease in the concentration of Na with the increase in Mg after treatment. Moreover, the pH of AG increased after treatment. The efficiency for pH slightly increased at first then remained steady with time (Fig. 5(a)).

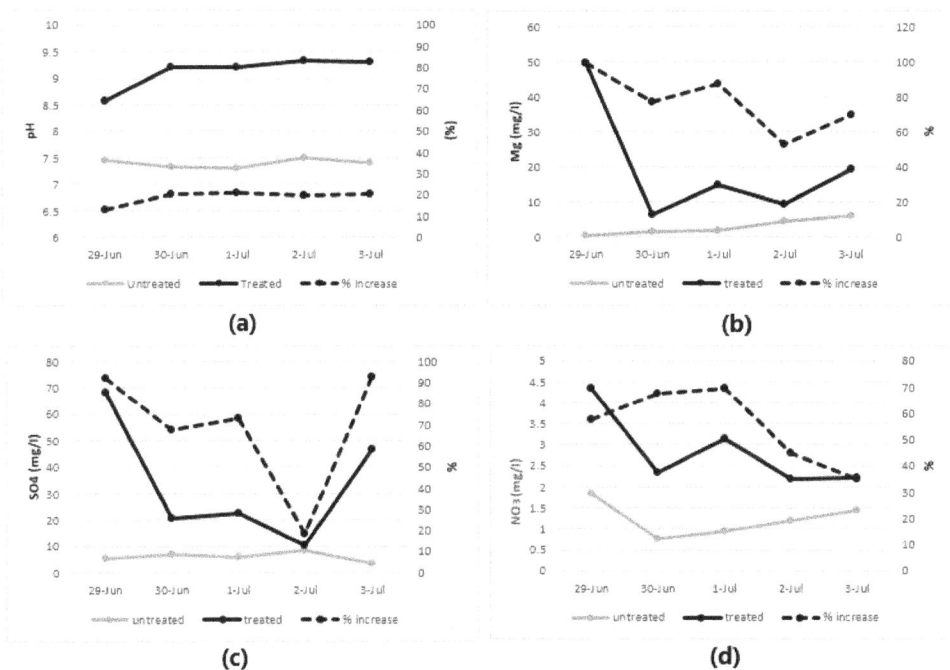

Figure 5: The concentration and removal efficiencies of (a) Mg; (b) SO_4^{2-}; (c) NO_3; and (d) pH in untreated and treated AG.

Usually the efficiency of greywater treatment system is assessed by their removal efficiencies of chemical and microbial contaminants that are of environmental and health concerns, and whether the treated greywater's quality is in compliance with local and/or international standards. The multimedia filter used in this study proved to be efficient in treating AG produced at MEC mosque. Despite being decreased (TSS, turbidity, K, BOD_5, Na, and Cl) or increased (SO_4^{-2}, NO_3, and Mg) or fluctuated (Ca and TDS), the concentration of chemical contaminants in treated AG remained far below the acceptable limits set by Omani standards for reusing wastewater. However, the pH slightly exceeded the permissible limit and, therefore needs to be adjusted before reusing the treated water for irrigation and toilet flushing purposes.

3.4 Cost of treatment unit

The capital cost and operation and maintenance cost of the treatment system are presented in Table 3. The total cost of setting up the designed filter is 309 OMR (approximately 803 USD) which is much less than the commercially available greywater treatment systems, as the approximate cost of treatment system with similar capacity (3 m³/d) is 9000 USD [9]. Whereas the O&M cost of the system is 80 OMR (approximately 208 USD). The reason behind the low cost is the way the treatment system is designed and constructed. The treatment system is specifically designed for the AG at MEC mosque, targeting the quality and quantity of AG produced. Moreover, the use of locally available materials such as sand, activated carbon and water storage tanks further reduced the cost of treatment system.

Table 3: Capital, operational and maintenance cost of treatment unit.

Items	Estimated cost (OMR)
Capital cost	
Fabrication of filter, plumbing and labour	120
Filter material	45
Two water pump	64
Two water storage tank	80
Total capital cost	309
Operational and maintenance cost	
Electricity per year	20
Maintenance per year	60
Total O&M cost	80

3.5 Water savings and financial benefits

Usually the amount of water saving resulted by using a greywater treatment system are not equal to the amount of greywater collected or treated, rather they are equal to the amount of water savings achieved by the user [19]. However, this study intends to use the treated greywater for toilet flushing and irrigation at MEC College where more than 5000 student study, hence huge amount of water goes into toilet flushing. Therefore, this study calculates the water savings as the amount of greywater treated, because it is assumed as per the observation that treated greywater produced will be completely reused daily.

The quantity of AG generated at the MEC mosque was estimated by monitoring the quantity of water used by per person per ablution and the number of worshippers per week [4]. The average quantity of water required by a person for performing a single ablution was found to be 6 litres by monitoring a number of worshippers at the mosque at several prayer times in a day. Based on the information obtained from the survey, the AG produced for the whole month is estimated as:

Number of worshippers per week × 4 weeks × volume of ablution water used by per person
= (2625 persons/week) × (4 weeks) × (6 litres/person)
= 63,000 litres/month = 63 m^3/month
= 16642.8 gallons/month

The financial benefit for the designed filter is calculated as follows [19]:

(Volume of water saving × Volumetric water rate) – O&M cost/year
= [(16642.8 × 12) × (0.003)] – 80
= 519.14 OMR/year
*According to Public Authority of Electricity and Water, the price for one gallon of water in Oman is 0.003 baisa.

Therefore, it can be seen that the designed treatment system for MEC mosque can save 63 m^3 of water per month and can also provide financial benefit by saving 519.14 OMR (approximately 1348.48 USD) yearly. Recycling AG at each mosque can bring immense long-term water conservation benefits and cost savings, especially when the freshwater reserves are depleting.

4 CONCLUSION

Ablution greywater (AG) being less contaminated provides an alternative source of water for irrigation and toilet flushing purposes. Results show that the quality of AG in terms of pH, BOD$_5$, TDS, NO$_3$, Cl, Na, Mg, K, Ca, SO$_4^{-2}$, and *E. coli* was acceptable for reuse as per the Omani standards, however the TSS and turbidity exceeded the permissible limits. After comparing the quality of AG with the past studies in Oman, this study concluded that the characteristics of AG significantly change with time and, therefore it is necessary to continuously monitor the quality of greywater and update treatment systems with time. The designed treatment system was effective in treating AG in Oman for irrigation and toilet flushing purposes. The concentration of chemical contaminants in treated AG remained far below the acceptable limits set by Omani standards for reuse of wastewater. However, the pH slightly exceeded the permissible limit and, therefore needs to be adjusted before reusing the treated water for irrigation and toilet flushing purposes. Furthermore, as per WHO guidelines, treated AG quality in this study is suitable for irrigation of fruit trees, fodder crops, and ornamentals. Lastly, this study concluded that the designed treatment system could save 63 m^3 of water monthly at MEC mosque, and can provide a yearly financial benefit of 519.14 OMR (approximately 1348.48 USD).

ACKNOWLEDGEMENT
The authors are acknowledging the financial support given by The Research Council (TRC) Oman under the Open Research Grant (ORG) Program to conduct this research.

REFERENCES
[1]   Ahmed, M., Prathapar, S. & Al-Abri, A., Guidelines for reuse of greywater in Oman: A proposal. *Second Oman-Japan Joint Symposium: Preservation of Environmental and Water Resources amid Economic Development*, Muscat, Oman, pp. 75–79, 2005.

[2]   Noutsopoulos, C. et al., Greywater characterization and loadings: Physicochemical treatment to promote onsite reuse. *Journal of Environmental Management*, **216**, pp. 337–346, 2018.

[3]   Alharbi, S., Shafiquzzaman, M., Haider, H., AlSaleem, S. & Ghumman, A., Treatment of ablution greywater for recycling by alum coagulation and activated carbon adsorption. *Arabian Journal for Science and Engineering*, 2019.

[4]   Suratkon, A., Chan, C. & Ab Rahman, T., SmartWUDHU': Recycling ablution water for sustainable living in Malaysia. *Journal of Sustainable Development*, 7(6), pp. 150–157, 2014.

[5]   Radin Mohamed, R., Adnan, M., Mohamed, M. & Mohd Kassim, A., Conventional water filter (sand and gravel) for ablution water treatment, reuse potential, and its water savings. *Journal of Sustainable Development*, 9(1), p. 35, 2016.

[6]   Al-Wabel, M., Simple system for handling and reuse of graywater resulted from ablution in mosques of Riyadh City, Saudi Arabia. *International Conference on Environment Science and Engineering*, Singapore, pp. 42–45, 2011.

[7]   Abu-Rizaiza, O., Ablution water: Prospects for reuse in flushing of toilets at mosques, schools, and offices in Saudi Arabia. *Journal of King Abdulaziz University-Engineering Sciences*, 14(2), pp. 3–28, 2002.

[8]   Al Mamun, A., Muyibi, S. & Abdul Razak, N., Treatment of used ablution water from IIUM Masjid for reuse. *Advances in Environmental Biology*, 8(30), pp. 558–564, 2014.

[9]   Prathapar, S., Ahmed, M., Al Adawi, S. & Al Sidiari, S., Design, construction and evaluation of an ablution water treatment unit in Oman: A case study. *International Journal of Environmental Studies*, **63**(3), pp. 283–292, 2006.

[10]   Al-Zu'bi, Y., Ammari, T., Al-Balawneh, A., Al-Dabbas, M., Ta'any, R. & Abu-Harb, R., Ablution greywater treatment with the modified re-circulated vertical flow bioreactor for landscape irrigation. *Desalination and Water Treatment*, **54**(1), pp. 59–68, 2015.

[11]   Jeong, H., Kim, H. & Jang, T., Irrigation water quality standards for indirect wastewater reuse in agriculture: A contribution toward sustainable wastewater reuse in South Korea. *Water*, **8**(4), pp. 169, 2016.

[12]   Prathapar, S., Ahmed, M., Al-Adawi, S. & Al-Sidiari, S., Variation in quality and quantity of greywater produced at two mosques in Oman. *International Wastewater Conference*, Salalah, Oman, 2004.

[13]   Al Barwani, A. & Helmi, T., Sea water intrusion in a coastal aquifer: A case study for the area between Seeb and Suwaiq, Sultanate of Oman. *Agricultural and Marine Sciences*, **11**, pp. 55–69, 2006.

[14]   Prathapar, S., Jamrah, A., Ahmed, M., Al Adawi, S., Al Sidairi, S. & Al Harassi, A., Overcoming constraints in treated greywater reuse in Oman. *Desalination*, **186**(1–3), pp. 177–186, 2005.

[15]   Duqm, Water quality protection technical note. Duqm, 2018.

[16]   Brown, C., Activated carbon | pureflow, *Pureflow*. https://www.pureflowinc.com/activated-carbon/. Accessed on: 11 Jun. 2019.

[17]   Fosso-Kankeu, E., Heever, C., Gericke, G., Lemmer, N. & Waanders, F., Evaluation of the performance of an activated carbon supplemented sand filter for the reduction of COD in brewery wastewater. *9th International Conference on Advances in Science, Engineering, Technology and Waste Management*, 2017.

[18]   Mallongi, A. et al., Clean water treatment technology with an up-flow slow sand filtration system from a well water source in the Tallo district of Makassar. *Journal of Environmental Science and Technology*, **10**(1), pp. 44–48, 2016.

[19] Gauley, B., Water savings and financial benefits of single-family package greywater systems. *Alliance for Water Efficiency*, Chicago, Illinois, 2017.

DEGRADATION OF BIODEGRADABLE SINGLE-USE PLATES AND WASTE BAGS IN TERRESTRIAL AND MARINE ENVIRONMENTS

KAIRE TORN, GEORG MARTIN & GRETA REISALU
Estonian Marine Institute, University of Tartu, Estonia

ABSTRACT

Littering and microplastic are a rising problem in both the terrestrial and marine environment. The degradation of macrolitter into smaller particles is driven by several factors such as: material, temperature, UV light, grazers, etc. Field exposure experiments were conducted within the natural systems to assess how quickly fragmentation occurs under different environmental conditions. The degradation of disposable paper plates, biodegradable plates, and biodegradable waste bags was studied in two terrestrial (open-air vs. buried in soil) and two marine (submerged in seawater above sediment vs. buried in sediment) conditions. Additionally, both biodegradable items were labelled as compostable by the producer. Degradation under natural condition did not meet established official standards for the biodegradation. Products degradation rates were highest in the marine environment for sample products that were submerged in seawater but above the sediment, followed by those that were in the terrestrial environment and buried in the soil. The rate of degradation was affected by the prevalence of grazers in combination with wave action for the litter submerged in seawater but above the sediment. The biodegradable plates were completely degraded under these two conditions within 4 months. The loss of material was highest during the first months. Slowest degradation occurred in the open-air where the loss of all materials was only 8–17% after 10 months. Biodegradable plastic bags decomposed remarkably only in the seawater above the sediment, in other environments the loss of weight was less than 20% during the period of 10 months.
Keywords: discarded macrolitter, plastic pollution, biodegradable products, plastic deterioration, fragmentation, sea water, natural environment.

1 INTRODUCTION

Marine litter is recognized as a growing global problem. Although, plastic recycling is increasing, millions of tonnes of plastic still reach the marine environment each year. Plastics are of increasing concern due to their persistence and effects on the ocean's wildlife and potentially humans [1]. In the European Union, 80 to 85% of marine litter (measured as beach litter counts) is plastic, with single-use plastic items representing 50% of this [2]. Single-use plastic is a challenge to manage due to its lengthy degradation process and the production of fragmented residues. In this context, the use of biodegradable polymers presents itself as an alternative for reducing the long-term impact plastic materials have on the environment.

The global production and use of biobased and/or biodegradable plastics is on the rise [3]. Biodegradable or compostable plastics can be produced from either bio-based or fossil raw materials. Biodegradation is a complex phenomenon which occurs in multiple steps: abiotic- and biotic-deterioration, biofragmentation, and microbial assimilation and mineralization [4]. Biodegradable plastic is designed to breakdown once immersed in soil or fresh water. Compostable materials are designed for compostability in terrestrial landscapes, typically that of an industrial composting facility [5]. In less suitable conditions, they might biodegrade slowly or not at all or even fragment into microplastics [6], [7]. Within the natural environment, several conditions may increase the effect of mechanical degradation such as environmental factors like exposure to UV radiation and mechanical abrasion by wave action, and so forth [8].

WIT Transactions on Ecology and the Environment, Vol 251, © 2021 WIT Press
www.witpress.com, ISSN 1743-3541 (on-line)
doi:10.2495/WS210051

The marine environment is not an optimal media for biodegradation as the necessary conditions for full degradation are not met (e.g. low temperature and oxygen level) [9]. At present, studies evaluating degradation of macrolitter in natural habitats [10], especially in marine environments are limited [11]. Degradation under laboratory conditions is different from degradation under natural conditions [12]. Laboratory tests are more suitable for assessing the stability of a material and do not take into account the influence of environmental factors. The aim of this study was to investigate the degradability of disposable plates and plastic bags in both the terrestrial and marine environment. The experiment was conducted in the natural environment to better examine the fragmentation of disposable items in different real-world conditions.

2   MATERIAL AND METHODS

2.1 Materials

Three widely used single-use products were selected for this study. Of these products, two types of plate (15 cm in diameter) were assessed. The first type of paper plate is one commonly used and composed of primarily of cellulose with a thin layer (less than 10% from product) of polyethene. The second type of plate selected is made of natural raw material and labelled as 100% biodegradable and compostable. The third item selected was a biodegradable waste bag with handles (6l volume) labelled as 100% biodegradable and compostable marked to conform with EN13432 standard. EN13432 requires compostable plastics to disintegrate after 12 weeks and completely biodegrade after six months under the conditions of industrial composting. As such ≥90% of the plastic material should have been converted to CO_2 with the remaining share degraded into water and biomass.

2.2 Experimental design

To contain the sample products, bags were constructed of fiberglass mesh (mesh size 4×4 mm). The selected mesh size allowed for fauna to access the investigated products. Prior to the experiment the dry weight of each sample products was measured. The products under investigation were each placed into their own mesh bag and labelled before being closed by sewing. Triplicate samples for each three products were used to investigate each of the environmental conditions.

The degradation rate of the assessed products was tested in two terrestrial (open-air, buried in soil) and two marine (submerged in seawater above sediment, buried in sediment submerged by seawater) environments. Sample bags were anchored in place for each of their associated environmental conditions. The buried samples were covered with a 5–10 cm thick layer of soil or sediment.

The samples were removed from the experimental sites 2, 3, 4, and 10 months after installation. The samples were packed separately and transported into lab for further analysis. The material was removed from the bag, cleaned gently and carefully from extras (soil, sediment, roots) and dried at room temperature to constant weight. Weight loss was used to quantify the extent of degradation. Results were expressed as the percentage of remained dry weight of material compared to initial weight.

2.3 Experimental location and conditions

In situ degradation experiment was carried out in June 2018–April 2019 in Saaremaa island, Estonia. Terrestrial setup was performed in mowed lawn in Kõiguste (58.3733°N,

22.9818°E) where humus rich garden soil prevails. Field experiments in marine conditions were conducted in shallow, semi-enclosed Kõiguste Bay, northern Gulf of Riga, NE Baltic Sea (58.3714°N, 22.9816°E). Bottom substrate of the location was fine sand with mud, water depth was 0.5 m. The distance between terrestrial and marine experimental plot was 250 m.

Experimental area lies in the northern part of the temperate climate zone, and in the transition zone between maritime and continental climate. Estonia has four seasons of near-equal length. For air temperature and sunshine hours nearest State Weather Service station (Roomassaare) was used. Water temperature was measured with a General Oceanics thermologger in Kõiguste Bay experimental site (Table 1). Ice coverage (thickness 5–15 cm) were registered in Kõiguste Bay (State Weather Service) in winter 2018–2019. During the experiment, the salinity of the bay was 5.5–5.8 PSU.

Table 1:   Range of monthly average of temperatures and sunshine hours from June 2018 to April 2019. Recorded maximum of temperature is given in brackets.

Season	Air temperature (°C)	Water temperature (°C)	Sunshine hours
Summer	15.2–20.3 (30.3)	15.9–23 (30.5)	286–377
Autumn	4.8–14.7 (22.5)	10.3–15.5 (21.5)	28–204
Winter	−1.9–0.9 (7)	NA	14–90
Spring	1.6–6.7 (22.5)	NA	135–319

2.4 Statistical analysis

The results of the field experiments were statistically analysed using the factorial ANOVA: tree type of litter items, duration of experiment, four different environmental conditions and their combinations as the independent variables, and loss of weight as the dependent variable. Effects were considered to be statistically significant at $p < 0.05$. Statistical analysis carried out in STATISTICA 10.

3 RESULTS

The experiment lasted for a period of 300 days beginning in the summer and concluding in the following spring. In general, the observed degradation was faster during the first months. We detected the decomposition of some of the sample products after as little as 3 months. The average degradation curves are plotted in Fig. 1.

The results of the experiment show that a products degradation rate varies as a factor of both its material composition and the environment in which it is immersed (Table 2, Fig. 2). Across the entirety of the experiment the change of mass was observed to be greatest when a product was submerged in seawater but above sediment, followed by that of being buried in soil. Submerged material was found to be completely or near totally fragmented after four months. Slowest degradation rate observed was in the open-air environment where the loss of all material was minor during the first 4 months, and only 8–17% by the conclusion of the experiment.

The weight loss of the biodegradable bags was significantly different from the plates and was found to be low (7–20% loss) in all environments except when above marine sediment (86% loss). The degradation of both paper and biodegradable plates was similar, except when plates were buried in soil.

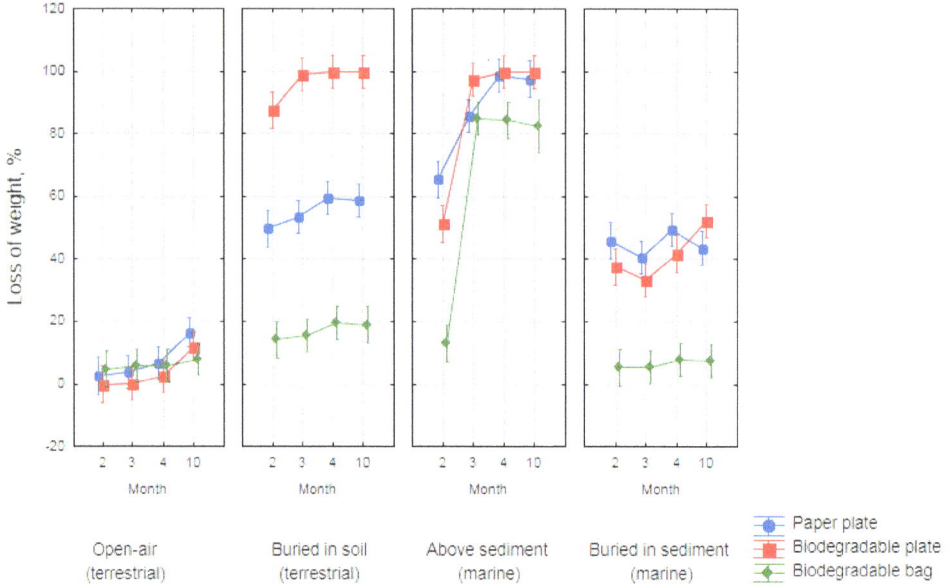

Figure 1:   Weight loss (mean ± 0.95 confidence interval, n = 3) of selected products in different environmental conditions.

Table 2: Results of factorial ANOVA for product material, duration and environment on the degradation of sample products. All effects are significant ($p < 0.05$).

	SS	Df	MS	F	p
Intercept	400,457	1	400,457	9,481.106	0.0000
Material	42,885	2	21,443	507.666	0.0000
Duration	9,490	3	3,163	74.895	0.0000
Environment	161,311	3	53,770	1,273.052	0.0000
Material*Duration	1,610	6	268	6.353	0.0000
Material*Environment	36,779	6	6,130	145.129	0.0000
Duration*Environment	14,759	9	1,640	38.827	0.0000
Material*Duration*Environment	3,530	18	196	4.643	0.0000

The biggest difference in the degradation rates for all of the measured products was when products were buried in the ground compared to when they were exposed in the open-air. If buried in the soil, biodegradable plates were completely fragmented after 3 months. In contrast, 41% of the paper plates and 11% of the biodegradable bags were remained at the end of experiment. Based upon visual estimations the biodegradable bags seemed usable with no fragmentation after a period of 10 months for the conditions of open-air and buried in sediment under seawater.

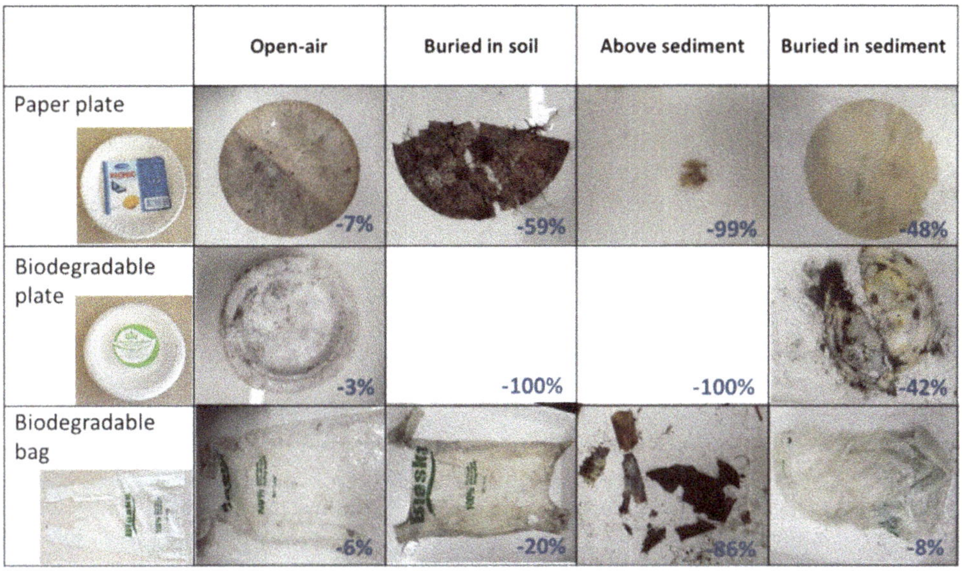

Figure 2:   Examples of products after a 4 month degradation period. Average proportion of weight loss is shown numerically.

4 DISCUSSION

The present study provides insight into the potential behaviour of paper plates, biodegradable plates, and biodegradable bags under different terrestrial and marine conditions over a period of 300 days. The results of the experiment showed that degradation rates varied as a consequence of a products material composition and the environment it is immersed in. In general, degradation was faster during the first few months.

The rate of biodegradation is dependent on the concentration of enzymes, microorganisms, temperature, pH value, humidity, oxygen supply, and light [6]. All these factors vary in different environments. In marine conditions, mechanical effects due to wave action are added. However, the impact of grazing invertebrates cannot be underestimated. Based on visual observations, items were inhabited by amphipods (Gammaridae, Corophidae) and isopods (Idoteidae) during sampling. Similar findings are reported by previous whereby numerous grazing crustaceans have been involved in the degradation of plastic items within the marine environment [13], [14].

Compost incubation is generally considered the most efficient method for biodegradation [15]. However, a great number of single-use products are released into the natural environment instead of being composted. Based on our results, degradation rates were highest in the marine environment for sample products that were submerged in seawater but above the sediment. It has been noted previously that plastics buried in marine environment take a long time for degradation to occur [16]. In the context of this study, the lowest rate of degradation was observed in open-air conditions for all four of the studied environments. Contrary to that of previous studies, the non-degradable plastic fragmented in open-air faster when compared to the marine environment [7], [17]. However, in these previous studies the impact of grazers was limited as the sampled products were shielded by dense mesh. Furthermore, by positioning samples higher in the water column and further from the sediment, the consequence was reduced degradation through impact by macrobenthos grazing. Additionally, the effect of UV-light and wave activity may have been greater in the present study due to shallow depth of the experimental site.

Different plastics display different characteristics depending upon their chemical composition. Surprisingly, the degradation of the biodegradable bags was remarkably low compared to the other materials investigated. The only environment where these bags lost more than 80% of their material when submerged in seawater above the sediment. The material of the bag directly relates to its decomposition rate as demonstrated by a previous experiment investigating compostable plastic bags which observed total decomposition in seawater within 3–5 months [11]. In contrast, biodegradable plates buried in the soil and in the sea above the sediment displayed the greatest loss of material, decomposing completely within 3–4 months. Additionally, standard paper plates displayed a higher rate of degradation when compared to the biodegradable bags.

Currently, there are a limited number of products capable of biodegradation under natural conditions or compostable by a typical household. Biodegradation tests that meet the European Union standards are performed under idealized conditions that can only be achieved by industrial biodegradation or composting operations. For example, when testing the compostability of a product, the temperature is kept constant at 58°C with humidity and aeration controlled at optimal levels [18]. In regard to biodegradation standards in seawater, water temperature is based upon a 30°C setting. In the northern seas, including the experimental area of the Baltic Sea, temperatures in coastal areas during summer typically reach 20–22°C with occurring 30°C under atypical weather conditions and only in shallow bays. Thus, it is generally not possible to meet these conditions for the degradation of biodegradable products in nature or at home. This likely results in confusion for consumers as the conditions of how the standards are set are not indicated on the packaging with the products claims of rapidly degrading in the environment most likely unachievable. Labels for disposable biodegradable products should include the information about the degradation condition as well as the material and additives used.

There is a common misconception that biodegradable plastics are safe and environmentally friendly because they are produced using natural materials. Several studies have expressed various concerns regarding biodegradable plastics. It is true that most biodegradable plastics are produced from materials of biological origin however the production process usually includes using different compounds including additives and plasticizers. It has been shown that in the European market 80% of bio-based and/or biodegradable plastic products contained more than 1,000 chemicals, with 67% of the products tested containing hazardous chemicals [19].

Member States of the European Union are required to ensure environmentally sound waste management to prevent and reduce marine litter from both sea and land sources. Therefore, in 2021 single-used plastic and oxo-degradable plastics will be banned in Europe. The objective of the European Strategy for Plastics is to ensure that all plastic packages in the Union market are re-usable or easily recycled [2]. The strategy of the European Commission focuses on recyclable plastics rather than biodegradable plastics. Biodegradable and/or compostable plastics require separate treatment through sorting of waste collection. There is also a risk that the use of biodegradable products may endorse littering behaviour.

5 CONCLUSION
The rate of a materials degradation in different environments will strongly depend upon the local conditions to which they are exposed. The degradation of biodegradable products under natural conditions do not meet the requirements for biodegradation defined by current EU

standards. The observed degradation rates were highest in the marine environment for sample products that were submerged in seawater but above the sediment, followed by those that were in the terrestrial environment and buried in the soil. In marine conditions, the impact of wave action and grazing by invertebrates on the degradation of a material cannot be underestimated. The degradation rate in an open-air was negligible. Biodegradable items from comprised of different materials showed very different results under the same environmental conditions. Therefore, labels for disposable biodegradable products should include the information about the appropriate degradation conditions as well as the materials and additives used.

ACKNOWLEDGEMENTS
This work was supported by the European Regional Development Fund via the Mobilitas Pluss (MOBERA12) of the Estonian Research Council. We would like to thank Imbi Püss and Tiia Möller for help in fieldwork. The authors thank Jack R. Hall for his help in editing the manuscript.

REFERENCES
[1]   Jambeck, J.R. et al., Plastic waste inputs from land into the ocean. *Science*, **347**(6223), pp. 768–771, 2015. DOI: 10.1126/science.1260352.
[2]   Directive (EU) 2019/904 of the European Parliament and of the Council of 5 June 2019 on the reduction of the impact of certain plastic products on the environment. *Official Journal of the European Union*, **L155**. https://eur-lex.europa.eu/legal-content/EN/TXT/PDF/?uri=CELEX:32019L0904&from=EN. Accessed on: 4 May 2021.
[3]   Report – Bioplastics Market Data 2019, Global production capacities of bioplastics 2019–2024. European Bioplastics, nova-Institute. https://docs.european-bioplastics.org/publications/market_data/Report_Bioplastics_Market_Data_2019.pdf. Accessed on: 4 May 2021.
[4]   Lucas, N., Bienaime, C., Belloy, C., Queneudec, M., Silvestre, F. & Nava-Saucedo, J.E., Polymer biodegradation: Mechanisms and estimation techniques – A review. *Chemosphere*, **73**(4), pp. 429–442, 2008. DOI: 10.1016/j.chemosphere.2008.06.064.
[5]   Biodegradable and compostable plastics – Challenges and opportunities. European Environment Agency, Briefing no. 09/2020. https://www.eea.europa.eu/publications/biodegradable-and-compostable-plastics. Accessed on: 4 May 2021.
[6]   Haider, T.P., Völker, C., Kramm, J., Landfester, K. & Wurm, F.R., Plastics of the future? The impact of biodegradable polymers on the environment and on society. *Angewandte Chemie International Edition*, **58**(1), pp. 50–62, 2019. DOI: 10.1002/anie.201805766.
[7]   Napper, I.E. & Thompson, R.C., Environmental deterioration of biodegradable, oxo-biodegradable, compostable, and conventional plastic carrier bags in the sea, soil, and open-air over a 3-year period. *Environmental Science and Technology*, **53**(9), pp. 4775–4783, 2019. DOI: 10.1021/acs.est.8b06984.
[8]   Kershaw, P.J., *Biodegradable Plastics and Marine Litter. Misconceptions, Concerns and Impacts on Marine Environments*. United Nations Environment Programme (UNEP): Nairobi, 2015.
[9]   Nazareth, M., Marques, M.R., Leite, M.C. & Castro, Í.B., Commercial plastics claiming biodegradable status: Is this also accurate for marine environments? *Journal of Hazardous Materials*, **366**, pp. 714–722, 2019.

[10] Green, D.S., Boots, B., Blockley, D.J., Rocha, C. & Thompson, R., Impacts of discarded plastic bags on marine assemblages and ecosystem functioning. *Environmental Science and Technology*, **49**(9), pp. 5380–5389, 2015.

[11] O'Brine, T. & Thompson, R.C., Degradation of plastic carrier bags in the marine environment. *Marine Pollution Bulletin*, **60**(12), pp. 2279–2283, 2010. DOI: 10.1016/j.marpolbul.2010.08.005.

[12] Rutkowska, M. et al., Biodegradation of modified poly (ε-caprolactone) in different environments. *Polish Journal of Environmental Studies*, **11**(4), pp. 413–420, 2002.

[13] Welden, N.A. & Cowie, P.R., Degradation of common polymer ropes in a sublittoral marine environment. *Marine Pollution Bulletin*, **118**(1–2), pp. 248–253, 2017.

[14] Hodgson, D.J., Bréchon, A.L. & Thompson, R.C., Ingestion and fragmentation of plastic carrier bags by the amphipod *Orchestia gammarellus*: Effects of plastic type and fouling load. *Marine Pollution Bulletin*, **127**, pp. 154–159, 2018.

[15] Roy, P.K., Hakkarainen, M., Varma, I.K. & Albertsson, A.C., Degradable polyethylene: Fantasy or reality. *Environmental Science and Technology*, **45**(10), pp. 4217–4227, 2011.

[16] Kumar, A.G., Anjana, K., Hinduja, M., Sujitha, K. & Dharani, G., Review on plastic wastes in marine environment – Biodegradation and biotechnological solutions. *Marine Pollution Bulletin*, **150**, p. 110733, 2020.

[17] Biber, N.F., Foggo, A. & Thompson, R.C., Characterising the deterioration of different plastics in air and seawater. *Marine Pollution Bulletin*, **141**, pp. 595–602, 2019.

[18] Innocenti, F.D., Biodegradability and compostability. *Biodegradable Polymers and Plastics*, eds. E. Chiellini & R. Solaro, Springer: Boston, MA, pp. 33–45, 2003. DOI: 10.1007/978-1-4419-9240-6_2.

[19] Zimmermann, L., Dombrowski, A., Völker, C. & Wagner, M., Are bioplastics and plant-based materials safer than conventional plastics? *In vitro* toxicity and chemical composition. *Environment International*, **145**, p. 106066, 2020. DOI: 10.1016/j.envint.2020.106066.

FOREST MANAGEMENT TO MITIGATE DISASTERS CAUSED BY HEAVY RAIN

KOJI TAMAI
Forestry and Forest Products Research Institute, Japan

ABSTRACT

Establishment of forest management methods to balance disaster mitigation function with timber production function has been sought in Japan. Forest maturity is thought to have improved its function to mitigate slope failures and flood disasters contributed by the reinforcing effect of slope stability by root system. It is ideal to avoid cutting down of forests with disaster occurrence risk. However, when cutting down of trees might be unavoidable for production of timber, harvested forest should be selected by disaster-occurrence risk and distance from residential area. Moreover, measures should be taken so that an increase in the disaster occurrence risk is suppressed to the greatest degree possible, such as suppression of the amount to be cut down and quick replanting after cut down.

Keywords: reinforcing effect of slope stability by root system, slope failure, reforestation.

1 INTRODUCTION

Precipitation amounts are high in Japan. Not uncommonly, hourly precipitation reaches as high as several hundred milli meters because of typhoons and Baiu front. Crustal movement is active in Japan. Therefore mountain slope stability is low. Slope failures occur more frequently during intense rainfall, resulting in numerous deaths. Therefore, people's expectations for forest functions to mitigate disasters are high. Because of timber resource shortages worldwide, the forest industry has remained active, even in Japan. Therefore, in Japan, establishment of forest management methods to balance disaster mitigation function with timber production function has been sought. This paper introduces forest management methods along with a review of earlier studies.

2 METHODOLOGY

By reviewing previous studies in Japan, it is shown that the slope failure prevention function and flood mitigation function of forests are derived from forest soils, and that the tree root system contributes greatly to the maintenance of forest soils. Based on the results of the review, we will discuss forest management that balances the functions of slope failure prevention, flood mitigation, and logging timber production.

3 RESULTS FROM THE REVIEW

3.1 Importance of forest soil in slope failure prevention and flood mitigation

Until the 1950s, woods used for building and fallen leaves used for fuel or fertilizer were supplied from forests. Wood and fallen leaves tended to be harvested more than the forest could supply. As a result, forests have become oligotrophic, declining and bare. Bare land was distributed throughout Japan at that time. In the 1950s, area of about 27,400 km^2 bare lands existed throughout Japan, whereas the forest area was about 250,800 km^2 (Chiba [1]). However now, bare land has been nearly eliminated. This transformation is attributable to enforcement of greenery work, as shown in Fig. 1 and to the fact that fallen leaves are not collected anymore because of the wider use of fossil fuels and chemical fertilizer.

WIT Transactions on Ecology and the Environment, Vol 251, © 2021 WIT Press
www.witpress.com, ISSN 1743-3541 (on-line)
doi:10.2495/WS210061

Figure 1:   Bare land in Japan before (left) and after (right) greenery work. In the greening work, saplings are planted along the contour lines of the terrace where the slopes have been cut out [9].

In addition to the fact that the bare land has almost disappeared, forests of Japan have matured since the 1960s, as represented by the fact that accumulation of trees has increased dramatically (Fig. 2). Before the 1950s, clear-cutting and reforestation have been done actively. However, after the 1960s, cutting down of trees has decreased due to stagnation of forest industry. New reforestation and young forest area also decreased, whereas areas of protected forests increased where logging was prohibited to control disasters.

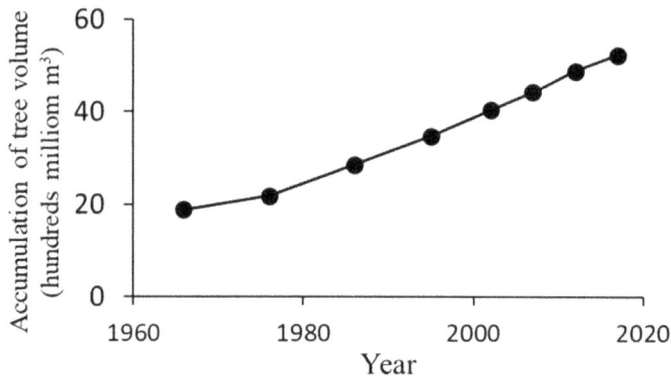

Figure 2: Transition of tree accumulation in Japan.

Keeping pace with the maturity of forests after the 1960s, the number of slope failure occurrences, areas damaged by flood water, and the number of deaths by heavy rain disaster decreased (Fig. 3). Suzuki [2] reported that slope failure occurring simultaneous frequently at almost all heads of streams during heavy rains were often observed before the 1980s. However, this sort of disaster is not observed since July 1983. Tamai [3] reports that the maturation of forests since the 1960s has improved the flood mitigation function of forests. This is because the development of tree roots suppressed the occurrence of slope failures and the forest soil developed.

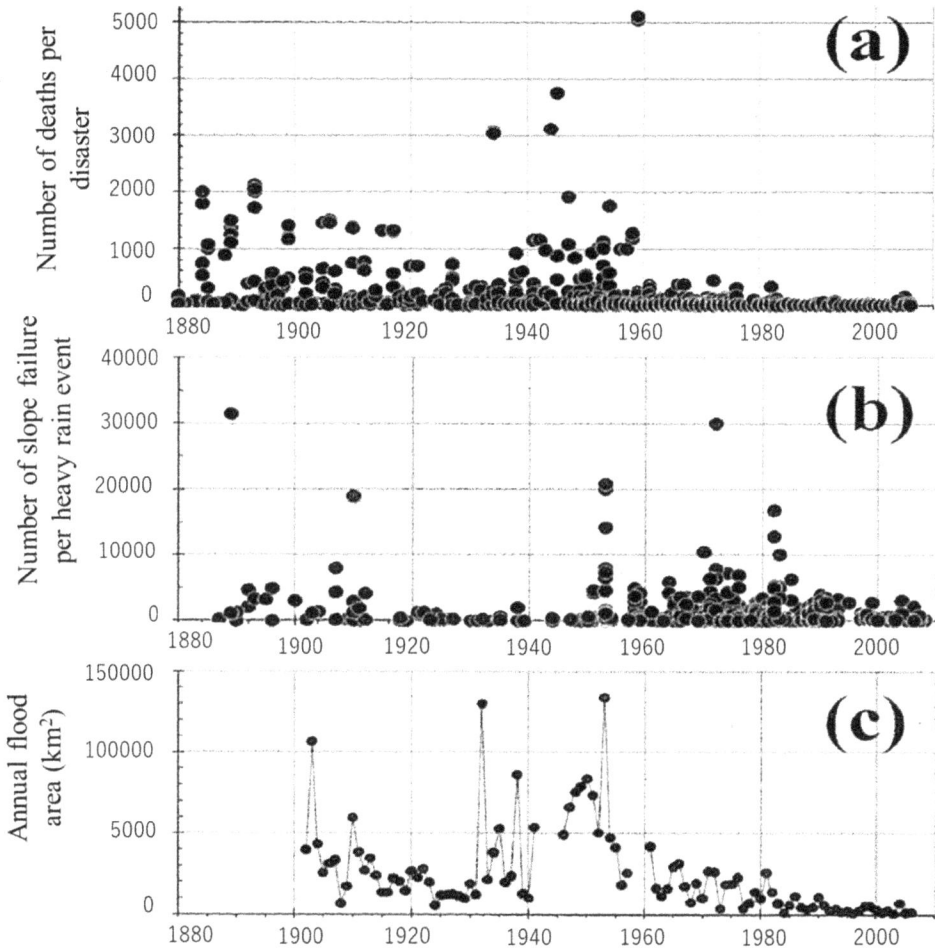

Figure 3:   Year-to-year fluctuations in disaster damage in Japan. (a) Deaths people number; (b) Slope failure number; and (c) Flood area. *(Source: Tada, 2018.)*

As mentioned earlier, forest maturity is thought to have improved its function to mitigate slope failures and flood disasters. Related mechanisms are explained in the next section.

3.2 Reinforcing effect of slope stability by root system

Roots of forest trees elongate as they entangle themselves among the soil particles. The roots of the tree then strengthen the connections between the soil particles and prevent slope failure during heavy rains. This effect is called "reinforcing effect of slope stability by root system". For this effect, if forests are clear-cut, then the strengthen effects of soil connections by roots decrease as the roots of a stump decay. The reinforcing effect by tree roots replanted immediately after clear-cutting increases as it grows. Kitamura and Namba [4] evaluated changes of effects by decaying stumps and effects by growing roots of replanted trees over the years since clear-cutting and reforestation (Fig. 4). The results clarified that it takes about

30 years before reinforcing effects by replanted tree roots become equal to reinforcing effects by tree roots before clear-cutting. It is also clearly recognized that reinforcing effects become minimal about 10–20 years after clear-cutting and reforestation. Tada [5] reported when slope failures occur as how many years after clear-cutting for two cases: reforestation is done after clear-cutting and the forest is left to natural recovery (Fig. 5). For natural recovery, the slope failure area continued to increase, whereas the number of years from clear-cutting increases. When reforestation is done after clear-cutting, the slope failure area reached a peak in the 20th year after clear-cutting.

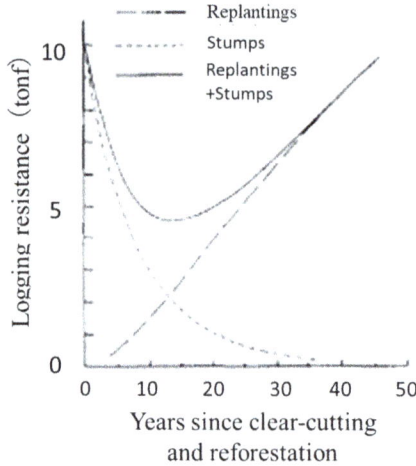

Figure 4: Changes in reinforcing effect of slope stability by root system over the years since clear-cutting and reforestation [4].

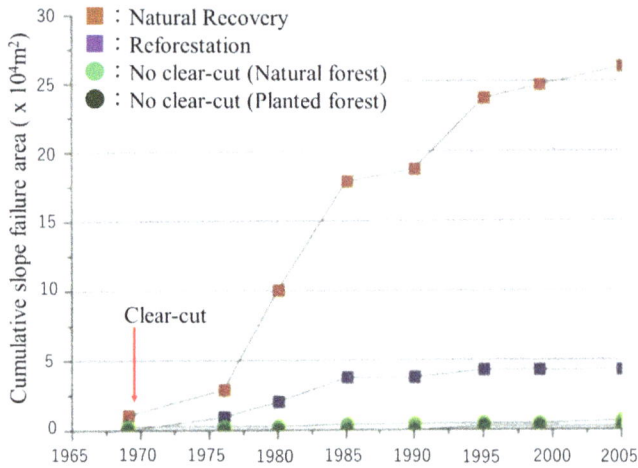

Figure 5: Difference in collapse area depending on the presence or absence of reforestation after clear-cutting [5].

For two adjoining forest basins where clear-cutting is made, Tamai [6] reported changes of the flood mitigation function of the forest as follows. Reforestation was made immediately after clear cutting in one basin. Results showed no reduction of flood mitigation function. In another basin, reforestation was not made immediately after clear-cutting. Subsequently, avalanche and slope failure occurred: about 20% of the basin area turned to bare area. The runoff delay effect of forest soil does not work on bare land (Tani et al. [7]). Therefore, it was observed that the flood water increased by 10%. This indicates that the reinforcing effect of slope stability by root system also contributes to the flood mitigation function of forest.

4 DISCUSSION TO ESTABLISH FOREST MANAGEMENT INCORPORATING DISASTER PREVENTION FUNCTIONS

In the preceding section, forest functions to mitigate slope failure and flood disaster by heavy rains are explained to be contributed by the reinforcing effect of slope stability by root system. In this section, forest management considering the effects of root system is explained based on reports by Tada [5], [8].

4.1 Concept of choosing a logging area from forests with disaster occurrence risk

Tada [8] defined three conditions for places where slope failure occurs easily: (1) steep slopes (more than 25 degree slope), (2) weathering of ground is advanced (soil layer is clay), and (3) ground water is concentrated. Geography designated as the head of a stream, fault topography, landslide morphology, geological boundary, and alluvial cones are cited.

In such places, the soil mass of the slope moves gradually downward over many years. A characteristic microrelief is formed in these areas (Fig. 6). On the upper end of the soil mass, a depression occurs first, which is enlarged gradually to a crack, and a cliff. At the lower end of the soil mass, an obduction, engulfment and embankment of moving of the soil mass are generated. While the soil mass moves downward gradually over time, trees grow on the soil mass. Japanese cedar and cypress, which are popular tree planting species in Japan, grow such that the trunk is kept vertical. However, for cedar and cypress grown on a soil mass that is moving gradually downward, because of soil mass deformation, an originally vertical stem will tilt. Cedar and cypress with a bent stem strive to return to vertical; for that reason, the stem is bent like a bow. For forests with many cedar and cypress with bent stem, the soil mass can be inferred as shifting. In other words, existence of cedar and cypress with a bent stem constitutes important information for inference that some risk of disaster occurrence exists.

For forests with disaster occurrence risk, slope failures should be prevented by reinforcing effect of slope stability by root system. Since the slope is stable due to the reinforcing effect of the root system, the flood mitigation function is also maintained. To do so, it is ideal to avoid cutting down of forests with disaster occurrence risk. However, cutting down of trees might be unavoidable for production of timber, even in forests with disaster occurrence risk. In such a case, two strategies can reduce the risk of disasters by choosing where forests can be cut and where they should be preserved. One can rank areas by the degree of disaster-occurrence risk. Places judged to have disaster occurrence risk are divisible into high risk places and low risk places. Forests with lower risk should be selected for cutting. Also, one can rank areas by the occurrence of disaster bringing human damage or economic damage. When a slope failure occurs at a place where there is residential areas and agricultural regions that is immediately threatened, human damage or economic damage can be caused. However, at a place distant from residential areas and agricultural regions, even if slope failure occurs, human damage and economic damage do not occur.

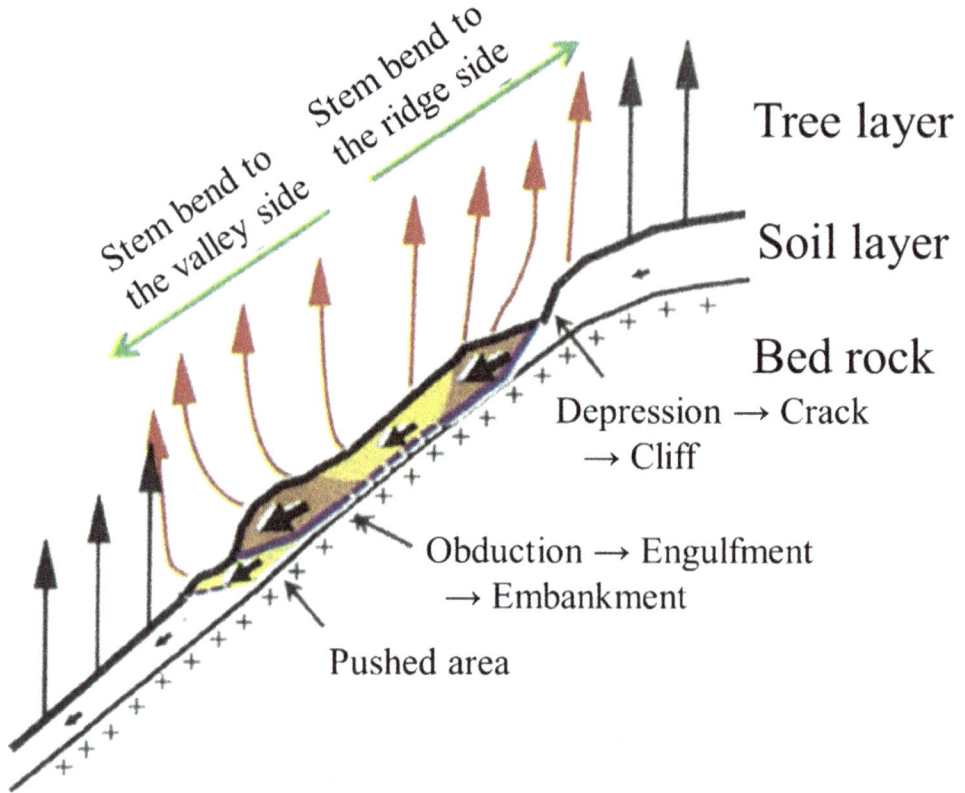

Figure 6:   Typical examples of the microtopography of forests with a high risk of slope failure and the bending of standing trees [8].

4.2 Forest measures to suppress disaster occurrence risk

When a forest should be cut down at the place judged to present disaster prevention risk, forest selection is explained in the preceding section. In this case, measures should be taken so that an increase in the disaster occurrence risk is suppressed to the greatest degree possible.

First, suppression of the amount to be cut down is considered. The tree roots are effective for suppressing slope failure. Fewer trees being cut down leads to less tree root reduction. Therefore, not all trees should be cut down. More trees left uncut leads to a greater suppression of the increase of disaster-occurrence risk. In other words, the rate of cutting should be kept as low as possible for the area presenting high disaster-occurrence risk, thereby maintaining high tree density.

Also, replanting should be conducted promptly after cutting. Fig. 3 shows that it takes 30 years before the tree root of a newly replanted young tree to grow and reinforcing effect of slope stability to be effective. Fig. 4 shows that slope failure can start after five years from cutting down because of the decay of stumps. For this reason, it is crucially important to perform replanting immediately after cutting. This replanting reduces the amount of time that the reinforcing effect of slope stability is weak.

5 CONCLUSION

The review showed that the forest slope failure prevention function and flood mitigation function are derived from forest soil, and that the tree root system contributes significantly to the conservation of forest soil.

Therefore, it is considered desirable for disaster mitigation to conserve the trees and maintain the root system. However, the timber production function is also one of the important functions of the forest.

When a forest should be cut down at the place judged to present disaster prevention risk, forests with lower risk or farther from the residential area is concluded to be selected for logging. In addition, it has also been concluded that limiting the number of loggings and prompt replanting after logging are essential.

REFERENCES

[1] Chiba, T., Bare land study in hilly mountains. *Sosiete*, p. 349, 1991. (In Japanese.)
[2] Suzuki, M., Influences of forest maturation on transition of sediment disasters. *SABO*, **100**, pp. 2–7, 2009. (In Japanese.)
[3] Tamai, K., The evaluation of forest functions of flood control and water resources conservation. *International Journal of Environmental Impact*, **3**, pp. 304–313, 2020.
[4] Kitamura, Y. & Namba, S., Tree roots upon landslide prevention presumed through the uprooting test. *Bulletin of the Forestry and Forest Products Research Institute*, **313**, pp. 175–208, 1981. (In Japanese.)
[5] Tada, Y., Aiming for both functions of forestry and national land conservation (2): Basic knowledge of disaster risk in forest area. *Sanrin*, **1641**, pp. 34–43, 2021. (In Japanese.)
[6] Tamai, K., Effects of timber cutting on maximum and minimum daily runoff during non-snow season: Cases in Kamabuchi and Tatsunokuchiyama experimental watersheds. *Journal of Japan Society of Hydrology and Water Resources*, **34**, in press. (In Japanese.)
[7] Tani, M. et al., Predicting the dependencies of rainfall-runoff responses on human forest disturbances with soil loss based on the runoff mechanisms in granite and sedimentary rock mountains. *Hydrological Processes*, **26**, pp. 809–826, 2012.
[8] Tada, Y., Aiming for both functions of forestry and national land conservation (1): Concept of forest disaster risk for forestry engineers. *Sanrin*, **1640**, pp. 37–45, 2021. (In Japanese.)
[9] Japan Forestry Agency. Greenery work in 1870–1910s, 2021. https://www.rinya.maff.go.jp/j/kouhou/archives/tisan/tisan.html.

HOW SOCIAL ARE FLOOD RISK MANAGEMENT PLANS IN SPAIN?

GUADALUPE ORTIZ[1], PABLO AZNAR-CRESPO[1] & ÁNGELA OLCINA-SALA[2]
[1]Department of Sociology I, University of Alicante, Spain
[2]Sciences Po, Paris School of International Affairs, France

ABSTRACT
Due to the weaknesses of the technocratic model in providing integrated, sustainable flood risk management, a new approach has emerged aimed at promoting adaptive strategies by means of non-structural measures, such as raising awareness among exposed populations and improving their adaptive response to risk. However, the emerging nature of these measures makes it necessary to analyse the way in which this social approach is being implemented in the day-to-day practice of flood risk management. Thus the aim of this paper is to assess the integration of social actions of risk response and public participation into risk management processes. To this end, 14 Spanish flood risk management plans (FRMPs) were reviewed and codified by means of a documentary content analysis. The coding process was designed according to the models of social capacity building and participatory assessment in flood risk management. The resulting analysis provided information on the characteristics of participatory processes in the development of FRMPs and on five areas of social initiatives for capacity building: knowledge, financing, motivation, governance and networks. The results revealed the limited quality of participatory processes due to procedural weaknesses and low representation of social actors. Also, social actions were focused on the promotion of risk awareness and showed a lack of technical detail that reduced their potential for implementation. Thus this paper argues that the social approach has not been transferred to practice in flood risk management, since the development of measures is still predominantly shaped by the traditional technocratic approach.
Keywords: disaster, natural hazard, social capacities, risk governance, non-structural measures, stakeholder, science-policy gap, social vulnerability.

1 INTRODUCTION
The technocratic paradigm prevailing in flood risk management has been called into question for its apparent limitations to provide sustainable and effective management [1], [2]. This paradigm is based on protection from and reaction to flooding through building defences aimed at draining floodwaters rapidly and effectively [3]. In line with this principle, flood risk analysis specialises in hazard studies that calculate the probabilities of periods of recurrence, model river courses in flood episodes and estimate direct material damages [4], [5]. In the area of management, the technocratic paradigm has led to the predominance of structural measures for retention, protection and drainage [6], [7], spurred by technological optimism and the belief in mastering nature [8]. These measures, since they are exclusively aimed at controlling the threat, have left aside factors such as land-use planning and its influence in increasing exposure to flood risk, the flow dynamics of historically floodable areas, and social vulnerability of the population [9]. This lack of attention to the whole set of forces driving risk creation has given rise to a reactive, deficient flood management model [10], [11] that has often been overwhelmed by the unforeseen size of events or by the population's high levels of exposure and vulnerability.

The inability of the technocratic model to create resilience to flood risk has favoured the appearance of new criteria and principles focusing more on adaptive management [12], [13]. This new approach calls for a transition from protection and reaction based on threat control, to prediction and adaptation based on strengthening the adaptive capacity of the risk exposure units [14]. The inclusion of adaptive criteria in management models makes it necessary to

WIT Transactions on Ecology and the Environment, Vol 251, © 2021 WIT Press
www.witpress.com, ISSN 1743-3541 (on-line)
doi:10.2495/WS210071

develop non-structural measures. These measures, rather than building physical defence facilities, are aimed at areas such as territorial planning and land use, the promotion of social awareness of risk and self-protection behaviours, the risk communication in the case of emergencies, institutional governance and legal reactions to disaster impacts [15]. The objective of such measures is not to mitigate or control the flood event but to reduce the exposure and vulnerability of the socio-territorial fabric as a whole [16]. Thus this new adaptive approach does not only involve analytical and technical changes in approaches to managing risk, but also represents a paradigm shift in the values mediating between the social, environmental and economic concepts of flood risk [17], [18].

One of the main paradigmatic consequences of this new management approach is its greater recognition of the social dimension of risk [6]. The idea that exposure units are proactive and can make adaptive responses in preparing for, confronting and recovering from risk obliges us to pay attention to the social processes behind such responses, particularly by analysing and managing social vulnerability [19]. This concept, the most frequently analysed variable in studies of the social dimension of flood risk [20], is defined as the sum of social, economic, political, institutional and cultural conditions determining the ability of people, groups and overall systems to tackle the negative impacts of stressful events and recover from the ensuing changes in the short, medium and long term [21]. The concept of vulnerability, then, goes beyond individual conditions of risk response, since it includes the forces creating it and enables the conceptualisation of the adaptive responses of systems as a whole. Owing to its integrated nature, social vulnerability can be seen not only as an operative concept or variable, but also as an overarching approach through which to understand the influence of the social structure in the creation of risk [22]. Due to its importance, studies of social vulnerability to flood risk have seen significant growth in recent years [23]. This research has mainly been aimed at assessing the socio-demographic conditions of exposure units [24], investigating the population's knowledge and perceptions of risk [25] and analysing models of socio-institutional governance of risk [26]. In the latter field of study, the emergence of social participation as a formula for proactive risk management is notable. Various analysts have indicated the need to enhance governance systems by expanding the community of social actors taking part in decision-making for the development and application of flood risk management measures [27]–[29]. Social participation in this field represents a means not only of democratising the development of management plans [30], but also of empowering the population by equipping it with the knowledge necessary for the effective application of responses to flooding [31].

Despite this recent growth in studies of the social dimension of flood risk, particularly through developing the social vulnerability approach, not enough evidence has yet been accumulated on how these new principles are being incorporated into the practice of risk management [32]. Some analysts have drawn attention to a science-policy gap in the field of flood risk management. Spray et al. [33] argue that risk managers and researchers have historically carried out their professional and scientific tasks in isolation from each other. According to these authors, risk managers have centred on factors for controlling the threat but have neglected to develop policies for land-use planning or reducing the population's social vulnerability. For this reason, management plans tend to lag behind contemporary knowledge and innovations in risk and disaster science [34]. Van Buuren et al. [13] argue that the technocratic tradition of risk managers, and political actors' need to satisfy public expectations of total protection through hydraulic defence works, are the two main factors behind the isolation of flood risk management and its difficulties in introducing innovative approaches that are sensitive to the social dimension.

Once this research problem had been defined, and with the purpose of contributing to the evidence available on the topic, this study's main objective was to analyse the extent to which the social dimension was encompassed in the practical management of flood risk in Spain. To this end a documentary content analysis of 14 Spanish flood risk management plans (FRMP) was performed, with the aim of assessing their integration of social actions for risk response and public participation. On the basis of the analysis of these documents, which are the main administrative instrument for flood risk planning in Spain, we attempted to ascertain whether and to what extent Spanish flood risk management is undergoing a shift towards new governance approaches.

2 METHODOLOGY

The object of our analysis, as we remarked above, was Spanish FRMPs; thus a sample of 14 approved plans for the different flood zones of the country were used, except that of Catalonia, which lacked homogeneity with the other plans. From each FRMP two documents were chosen for analysis: (a) the "Description of Strategy Programmes" annexe, in which the strategic lines of action aimed at enabling society to respond to the risk are described, and (b) the "Summary of Public Information and Consultation Processes and their Results" annexe. A documentary content analysis was chosen for the purpose. This qualitative analysis used both deductive coding (on the basis of previously defined conceptual categories drawn from the specialised literature) and applied inductive coding (through the creation of codes on the basis of an exhaustive review of the texts in the light of the research objectives and questions). The analysis was performed using the qualitative content analysis software Atlas.ti.

After reviewing the strategic lines described in each document, those with a marked social character were selected. These fell into 5 main types of social strategies: (a) the development of studies for improving knowledge of flood risk management; (b) strategies aimed at setting up or improving institutional planning of responses to flood emergencies through coordinating civil protection plans; (c) improvement of protocols for action and communication of information in cases of flooding; (d) strategies aimed at raising public awareness in preparation for floods, in order to increase flood risk perception and self-protection among the population and social and economic actors; and (e) civil protection plans: actions for supporting health, financial (including legal) assistance, and the temporary rehousing of affected populations. Each of these strategic lines was specified in turn in "Specific Actions", which were the object of the documentary content analysis.

As we mentioned above, the content analysis was approached in two different ways, corresponding to the two types of documents analysed: first, by analysing the particular actions specified in the different strategic lines; and second, by analysing the quality of the participatory strategies adopted in the process. To analyse the actions from the strategy programme documents, Kuhlicke et al.'s model [29] of social capacity building for natural hazards was used, in Ballester's version [35], adapted to flood risk management in Spain. This model comprises five dimensions: knowledge, motivation, networks, governance and financing. Thus, adopting the social vulnerability approach, we performed an analysis aimed at ascertaining whether these dimensions of capacity-building were reflected in the specific actions programmed in the strategic lines in each FRMP studied. Further, it was chosen to adopt an additional coding model, using the classical agency/social structure pairing. Agency was defined as the tendency of the actions to increase the affected population's individual responsibility and their empowerment to respond to flood risk. As for structure, this was defined as the institutional actions undertaken to create directives and compulsory procedures for social actors. This twofold categorisation enabled us to identify the direction of each of the actions making up the strategies.

The documents on public information and consultation were analysed using the model for assessing participation in flood risk management developed by Maskrey et al. [36]. This model assesses the quality of participatory processes in three main dimensions: context, process and outcomes, dividing the latter into substantive and social outcomes. Context indicates the predisposition of the socio-institutional fabric towards participation, embodied in factors such as understanding of the phenomenon of floods, institutional support, previous interactions among stakeholders and the complexity of disputes. Process criteria assess the common characteristics of the participatory processes that impact the efficacy of participation, and are divided into five subcategories: accessibility, deliberation, representation, response and quality. Lastly, social outcomes assesses community capacity-building to respond to floods and mitigate vulnerability through participatory processes, analysis of which is subject to the social content of the allegations presented by each participant. It should be noted, however, that the category of substantive outcomes was discarded. This criterion assessed stakeholders' perceptions and attitudes towards expectations, objectives and conflicts once the participatory process was finalised, and was ruled out because the documents studied did not provide this information.

3 RESULTS AND DISCUSSION

Our analysis of the results yielded numerous indicators providing answers to our question on the extent to which the social dimension and public participation are included in Spanish FRMPs. The review of actions in the strategy programmes and consultation processes revealed extreme weakness in the integration of the social dimension and merely superficial treatment when it was timidly included in the plans. Firstly, there was a high level of similarity among the social strategies in all the documents from the 14 hydrological regions analysed. This similarity was so strong that on most occasions the actions were described in the exact same words, which would seem to suggest that the social strategies were extracted from a generic model without the necessary adaptation to the particular social context of each case. Despite this, although most of the actions were repeated for all strategies in all the FRMPs, there were a few exceptions in cases that included additional actions formulated exclusively for three regions: the Tagus and West and East Cantabria. Thus the analysis showed very little depth in the social strategies, and an absence of contextualisation of the proposed actions in accordance to the different social situations and specific cultures of each flood zone.

Fig. 1 shows the number of actions in each of the five types of social strategies analysed according to their capacity-building features. The dimension of knowledge (136) was the most frequent, followed by that of networks (108), motivation (42), financing (13) and governance (2). Nevertheless, it should be noted that the sum of elements from all the dimensions of the social capacity-building model (301) did not correspond to the total number of social actions in the strategy documents, since some were classified in more than one capacity-building dimension. The total number of actions was 179; thus, bearing in mind that each FRMP put forward an average of 20 strategic lines and that only five of these were categorized as strategies with social content, the slight proportional weight of the social actions in the whole set of plans becomes evident.

Turning to the orientation of the actions (Fig. 2), it can be observed that the category of structure was noticeably more prominent than that of agency, thereby demonstrating the clear predominance of the socio-structural tendency in the strategic lines, as opposed to actions

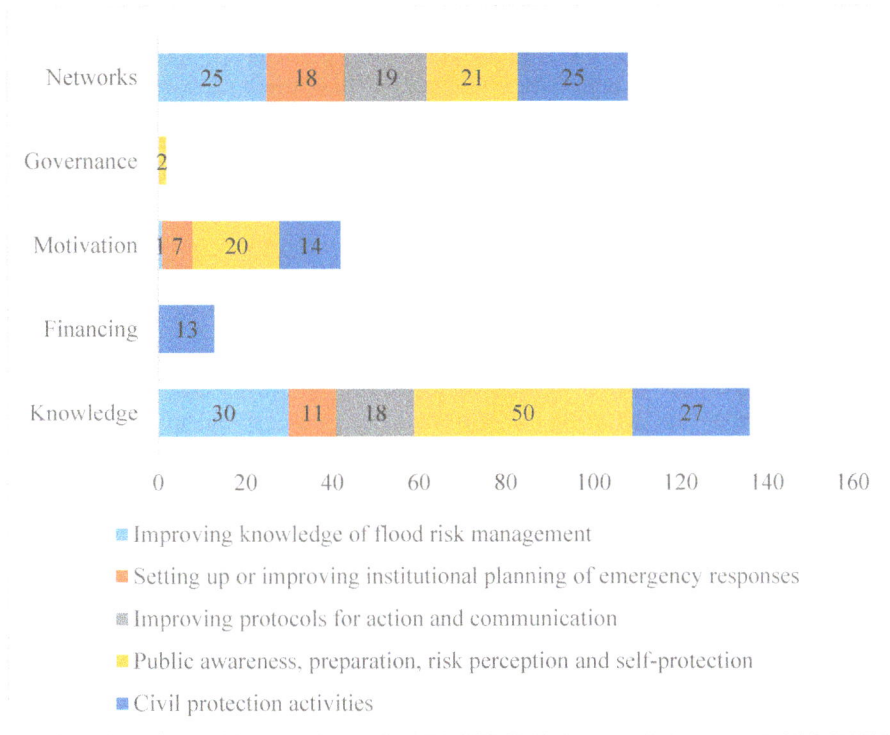

Figure 1: Number of actions according to social strategies and analytical categories. *(Source: Own elaboration.)*

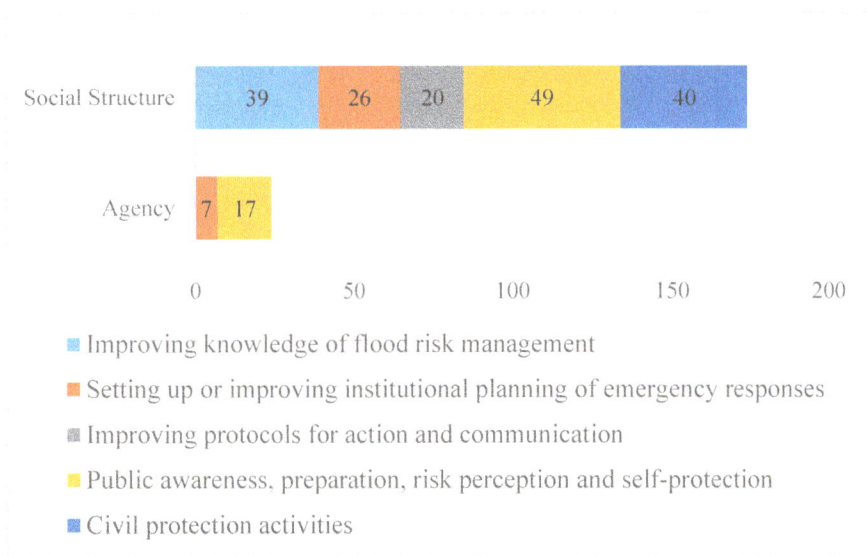

Figure 2: Number of actions according to social strategies and orientation (agency/ structure). *(Source: Own elaboration.)*

aimed at individual capacity-building for affected populations (who were thus assigned the role of passive receivers), including even strategies aimed at spreading the knowledge of risk.

Turning our attention to each of the five capacity-building dimensions, we observe that the dimension of knowledge was mainly specified through actions directed towards developing, improving and disseminating knowledge of flood risk and action protocols. These were institutional initiatives focussing on spreading information about self-protection and health guidelines among the population. Nevertheless, there was no observable direct engagement of the institutions themselves in ensuring a solid base of knowledge extending to the whole population, since only references to campaigns and websites appeared, with no actions requiring fluid communication with and direct involvement of the public. In the dimension of networks, the actions were directed towards greater coordination and cooperation among authorities and strengthening the system of actors through information campaigns. This gave increased weight to the socio-structural nature of the strategies, since the dimension revealed greater efforts towards building social capital among the authorities while, in contrast, only marginal attention was given to the networks of actors affected by floods. On the other hand, the actions in the dimension of motivation were closely linked to agency, particularly through the dissemination of guidelines for individual self-protective behaviour. The dimension of financing was, in itself, clear proof of the scarce integration of the social dimension in the strategic lines of the FRMPs: first, only one action of this type emerged, relating to aid and compensation for damages and repeated in exactly the same words in 13 of the 14 FRMPs analysed; and second, its scarce treatment, in terms of both frequency and development, indicated a lack of economic support assigned to carrying out those actions that went beyond merely technical work. The dimension of governance, the last and most underdeveloped category, only appeared in the FRMPs of the areas of West and East Cantabria, which contained actions referring to participatory workshops and questionnaires. In fact, the lack of the governance dimension in the actions also showed Spanish FRMPs to subscribe to a management model directed towards rigid structural measures based on top-down decision-making that did not facilitate collaborative relationships with the affected populations.

Regarding the inclusion of citizen participation strategies in the documents on public information and consultation, our analysis of the context and process categories (composed in turn of the subcategories of accessibility, deliberation, representation, responses and quality) evidenced the lack of detailed information in these categories, also heightened by the homogeneity and rigidity of the documents discussed above. For this reason, it was chosen to make a general descriptive analysis of these categories, aimed at developing an ideal type that would enable us to characterise the nature of public participation in Spanish FRMPs, rather than a systematic coding system and count of codes like that used in analysing the strategies and actions. In the subcategory of representation, on the other hand, it was possible to adopt a strategy of inductive coding, since the documents provided listings of social actors.

The information that the FRMPs provided on the contexts of their participatory processes was limited to mentioning the legislation, specifically Royal Decree 903, 9 July 2010, thereby once again demonstrating a severe lack of adjustment to the particular social and cultural characteristics of each region. Thus the only references to contextual issues were, paradoxically, very similar for all of the flood zones, with no real discussion of the contexts and social needs framing the processes, and with simple justifications indicating only the regulation to be complied with. In the dimension of process, our review revealed the weakness of participation in all five sub-categories. First, it can be concluded that accessibility in the participatory processes of the FRMPs was limited, since there were no discernible clear protocols for fostering and publicising participation, and the number of participants in some cases was low. Secondly, the level of deliberation was also low, since this was limited to the exchange of information between the interested party and the government, with no place for dialogue or communication with any other actors. The participatory process, consisting only of the circulation of official documents, allowed for intervention and response solely by the relevant authority, without creating spaces for debate or fluid exchanges of knowledge. In the subcategory of responses, participation in the FRMPs was framed in a rigid procedure restricted to mere application of the existing legislation, without offering the possibility of change through participatory action or adaptation to participants' objectives and timetables. Lastly, since the subcategory of quality referred to professionalism, clarity of objectives and participants' knowledge of what was expected of them, we can speak of a high level of quality, since the formality and the official-administrative nature of the documents ensured professional participatory processes with clear objectives. In general terms, however, we found participation in the FRMPs to be characterised by rigidity and superficiality, allowing hardly any exchange of knowledge or debate on the key issues. Moreover, the participatory processes were identical for all flood zones, and did not flexibilize or adapt their tools or procedures to the varying social and cultural situations of the different areas of Spain they dealt with.

Turning our attention to representation, as we can see from Fig. 3, the analysis of the public information and consultation documents and their results and the categorization of each of the social representatives participating in the consultation processes revealed a wide variety of actors, with a significantly greater number of companies, regional state bodies and town/city councils than others. Despite this, in our view, participative representation in the FRMPs was relatively low. Bearing in mind the size of each flood zone, many social actors on the local and infralocal scales were left out of account, particularly representatives of citizens' associations, non-state/non-company sectors and individuals with private interests.

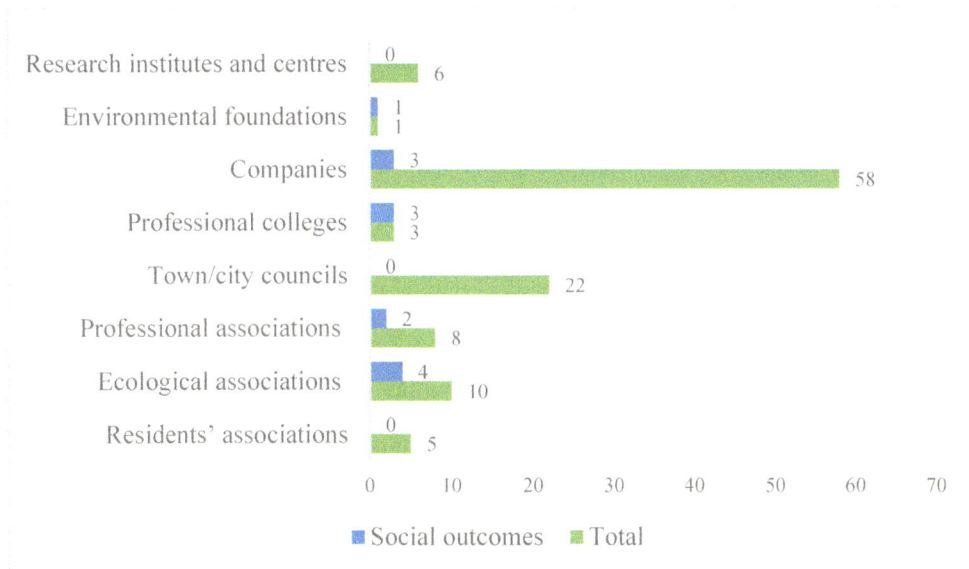

Figure 3:   Relative weight of social actors and number of responses to consultation on social outcomes from each actor. *(Source: Own elaboration.)*

Thus public bodies on the local, county, regional and national levels were, taken as a whole, the most prominent category. This may demonstrate that consultation processes were not clearly communicated to the public, and that the authorities did not promote greater citizen participation in the design and approval of FRMPs.

Further, it should be noted that in the category of social outcomes, this type of content was scarce compared to technical outcomes or results from structural measures. This deficiency is particularly striking among responses to consultation by state bodies representing the general social interest, such as town/city councils, regional and national governments and even political parties. Most responses to consultation in the area of social outcomes referred to the need to create greater awareness of flood risk and promote self-protection initiatives, i.e. measures aimed at enhancing the population's individual responsibility. There was virtually no mention of the active involvement of the public in truly participatory processes.

4 CONCLUSIONS

In terms of the main objective of this study, it can be concluded that Spanish FRMPs showed significant deficits in integrating the social dimension into the design of action strategies, since the social strategies found were marginal and extremely superficial. Participatory processes also showed severe weaknesses and did not effectively democratise public flood risk management, nor did they sufficiently embrace the social interests of the affected actors in their management strategies. In addition to this, our analysis revealed that attention to social content in the plans was reproduced in a standardised and decontextualised way for the majority of flood zones.

As noted in the preceding sections, this study used a social vulnerability approach to assess the integration of the social dimension into Spanish FRMPs. Our analyses in the areas of strategies and actions and the social results of participatory actions revealed the lack of a firm political commitment to strengthening and improving social capacity-building as a formula for boosting the population's adaptive responses and proactively addressing flood disaster risk. This finding was both clear and explicit, not only in terms of the very small number of social strategies included in the plans, but also in the notable weakness of their content. While these strategic lines seemed to be moving towards some kind of endeavour to overcome the traditional exclusive reliance on structural solutions, at the same time they had strikingly little materialisation in concrete, specific initiatives backed by economic, technical and social incentives. In contrast, the structural measures were sufficiently specific to be put into practice, thus evidencing the persistence and the unchallenged nature of the traditional technocratic paradigm, in which prevention, response and recovery measures are based mainly on technical solutions. Thus we may conclude that Spanish flood disaster risk management is a clear example of the science-policy gap [33], [34]. While scientific research has demonstrated the need to encompass the social dimension and public participation for more effective treatment of natural hazards, public management still reproduces the predominant technocratic paradigm, which has been shown to be insufficient. Thus in our view, the relationships between society and the environment are being recast on the academic and theoretical level, but this has not yet influenced the practical development and implementation of environmental policies. Although some leading international organisations and institutions, such as the European Environmental Agency [37] or the UNDRR [38], have called attention to the need to adopt more proactive and integrated approaches which would allow greater understanding and more effective tackling of socio-environmental risks, the inclusion of non-structural measures in local management systems and mechanisms is still vague, poorly defined and superficial. In short, addressing the question of whether the paradigm is currently shifting or persisting, we would argue that there is clear reproduction and continuance of the predominant technocratic paradigm, with only imprecise and marginal application of specific initiatives and strategies for putting into practice the scientific community's demands for new, more integrated models.

ACKNOWLEDGEMENT
This study was carried out within the framework of the National Programme for University Teacher Training (FPU) of the Spanish Universities Ministry and funded by a grant awarded to the second author of the paper.

REFERENCES
[1]   Pahl-Wostl, C., Becker, G., Knieper, C. & Sendzimir, J., How multilevel societal learning processes facilitate transformative change: A comparative case study analysis on flood management. *Ecology and Society*, **18**(4), 2013.
[2]   Schoeman, J., Allan, C. & Finlayson, C.M., A new paradigm for water? A comparative review of integrated, adaptive and ecosystem-based water management in the Anthropocene. *International Journal of Water Resources Development*, **30**(3), pp. 377–390, 2014.
[3]   Wolsink, M., River basin approach and integrated water management: Governance pitfalls for the Dutch Space-Water-Adjustment Management Principle. *Geoforum*, **37**(4), pp. 473–487, 2006.
[4]   Ayala-Carcedo, F.J., El sofisma de la imprevisibilidad de las inundaciones y la responsabilidad social de los expertos. Un análisis del caso español y sus alternativas. *Boletín de la Asociación de Geógrafos Españoles*, **33**, pp. 79–92, 2002.
[5]   Apel, H., Aronica, G.T., Kreibich, H. & Thieken, A.H., Flood risk analyses: How detailed do we need to be? *Natural Hazards*, **49**(1), pp. 79–98, 2009.
[6]   Brown, J.D. & Damery, S.L., Managing flood risk in the UK: Towards an integration of social and technical perspectives. *Transactions of the Institute of British Geographers*, **27**(4), pp. 412–426, 2002.
[7]   Few, R., Flooding, vulnerability and coping strategies: Local responses to a global threat. *Progress in Development Studies*, **3**(1), pp. 43–58, 2003.
[8]   Johnson, C.L. & Priest, S.J., Flood risk management in England: A changing landscape of risk responsibility? *International Journal of Water Resources Development*, **24**(4), pp. 513–525, 2008.
[9]   Samuels, P., Klijn, F. & Dijkman, J., An analysis of the current practice of policies on river flood risk management in different countries. Irrigation and Drainage. *The Journal of the International Commission on Irrigation and Drainage*, **55**(S1), S141–S150, 2006.
[10] Jeffers, J.M., Integrating vulnerability analysis and risk assessment in flood loss mitigation: An evaluation of barriers and challenges based on evidence from Ireland. *Applied Geography*, **37**, pp. 44–51, 2013.
[11] Driessen, P.P. et al., Governance strategies for improving flood resilience in the face of climate change. *Water*, **10**(11), p. 1595, 2018.
[12] Shrubsole, D., From structures to sustainability: A history of flood management strategies in Canada. *International Journal of Emergency Management*, **4**(2), pp. 183–196, 2007.
[13] Van Buuren, A., Lawrence, J., Potter, K. & Warner, J.F., Introducing adaptive flood risk management in England, New Zealand, and the Netherlands: The impact of administrative traditions. *Review of Policy Research*, **35**(6), pp. 907–929, 2018.

[14] Ward, P.J., Pauw, W.P., Van Buuren, M.W. & Marfai, M.A., Governance of flood risk management in a time of climate change: The cases of Jakarta and Rotterdam. *Environmental Politics*, **22**(3), pp. 518–536, 2013.

[15] Schanze, J. et al., *Systematisation, Evaluation and Context Conditions of Structural and Non-structural Measures for Flood Risk Reduction*, CRUE Funding Initiative on Flood Risk Management Research: London, 2008.

[16] Kundzewicz, Z.W., Non-structural flood protection and sustainability. *Water International*, **27**(1), pp. 3–13, 2002.

[17] Werritty, A., Sustainable flood management: Oxymoron or new paradigm? *Area*, **38**(1), pp. 16–23, 2006.

[18] Wiering, M. et al., Varieties of flood risk governance in Europe: How do countries respond to driving forces and what explains institutional change? *Global Environmental Change*, **44**, pp. 15–26, 2017.

[19] Fuchs, S., Karagiorgos, K., Kitikidou, K., Maris, F., Paparrizos, S. & Thaler, T., Flood risk perception and adaptation capacity: A contribution to the socio-hydrology debate. *Hydrology and Earth System Sciences*, **21**(6), pp. 3183–3198, 2017.

[20] Orimoloye, I.R., Belle, J.A. & Ololade, O.O., Exploring the emerging evolution trends of disaster risk reduction research: A global scenario. *International Journal of Environmental Science and Technology*, **18**, pp. 673–690, 2021.

[21] Birkmann, J. et al., Framing vulnerability, risk and societal responses: The MOVE framework. *Natural Hazards*, **67**(2), pp. 193–211, 2013.

[22] Singh, S.R., Eghdami, M.R. & Singh, S., The concept of social vulnerability: A review from disasters perspectives. *International Journal of Interdisciplinary and Multidisciplinary Studies*, **1**(6), pp. 71–82, 2014.

[23] Díez-Herrero, A. & Garrote, J., Flood risk analysis and assessment, applications and uncertainties: A bibliometric review. *Water*, **12**(7), pp. 2050–2074, 2020.

[24] Rufat, S., Tate, E., Burton, C.G. & Maroof, A.S., Social vulnerability to floods: Review of case studies and implications for measurement. *International Journal of Disaster Risk Reduction*, **14**, pp. 470–486, 2015.

[25] Raška, P., Flood risk perception in Central-Eastern European members states of the EU: A review. *Natural Hazards*, **79**(3), pp. 2163–2179, 2015.

[26] Morrison, A., Westbrook, C.J. & Noble, B.F., A review of the flood risk management governance and resilience literature. *Journal of Flood Risk Management*, **11**(3), pp. 291–304, 2018.

[27] Renn, O., *Risk Governance: Coping with Uncertainty in a Complex World*, Earthscan: London, 2008.

[28] Walker, G., Whittle, R., Medd, W. & Watson, N., Risk governance and natural hazards. CapHaz-Net WP2 Report, Lancaster Environment Centre, University of Lancaster, 2010.

[29] Kuhlicke, C. et al., Perspectives on social capacity building for natural hazards: Outlining an emerging field of research and practice in Europe. *Environmental Science and Policy*, **14**(7), pp. 804–814, 2011.

[30] Raymond, C.M., Fazey, I., Reed, M.S., Stringer, L.C., Robinson, G.M. & Evely, A.C., Integrating local and scientific knowledge for environmental management. *Journal of Environmental Management*, **91**(8), pp. 1766–1777, 2010.

[31] Hernández-Mora, N. & Ballester, A., Public participation and the role of social networks in the implementation of the Water Framework Directive in Spain. *Ambietalia*, **1**(Extra 1), pp. 1–21, 2011.

[32] Nordbeck, R., Löschner, L., Pelaez Jara, M. & Pregernig, M., Exploring science–policy interactions in a technical policy field: Climate change and flood risk management in Austria, Southern Germany, and Switzerland. *Water*, **11**(8), pp. 1675–1701, 2019.

[33] Spray, C., Ball, T. & Rouillard, J., Bridging the water law, policy, science interface: Flood risk management in Scotland. *Journal of Water Law*, **20**(2–3), pp. 165–174, 2009.

[34] Hegger, D., Alexander, M., Raadgever, T., Priest, S. & Bruzzone, S., Shaping flood risk governance through science-policy interfaces: Insights from England, France and the Netherlands. *Environmental Science and Policy*, **106**, pp. 157–165, 2020.

[35] Ballester, A., Participación pública para una gestión eficaz del riesgo por inundación: construcción de capacidades sociales en la Ribera Alta del Ebro (España). Doctoral thesis, Universitat Autònoma de Barcelona, 2017.

[36] Maskrey, S.A., Priest, S. & Mount, N.J., Towards evaluation criteria in participatory flood risk management. *Journal of Flood Risk Management*, **12**(2), pp. 1–14, 2019.

[37] European Environmental Agency, Disasters in Europe: More frequent and causing more damage.
https://www.eea.europa.eu/highlights/natural-hazards-and-technological-accidents.
Accessed on: 3 Apr. 2021.

[38] UNDRR, Sendai framework for disaster risk reduction 2015–2030.
https://www.undrr.org/publication/sendai-framework-disaster-risk-reduction-2015-2030. Accessed on: 3 Apr. 2021.

DESIGN OF A SEWAGE AND WASTEWATER TREATMENT SYSTEM FOR POLLUTION MITIGATION IN EL ROSARIO, EL EMPALME, ECUADOR

BETHY MERCHAN[1,2,3], PAULA ULLAURI[1,2], FERNANDO AMAYA[1,2], LENIN DENDER[1,2], PAUL CARRIÓN[1,2] & EDGAR BERREZUETA[4]
[1]Faculty of Engineering in Earth Sciences, FICT, Escuela Superior Politécnica del Litoral, ESPOL, Ecuador
[2]Center for Research and Projects Applied to Earth Sciences (CIPAT),
Escuela Superior Politécnica del Litoral, Ecuador
[3]Geo-Resources and Applications, GIGA, Escuela Superior Politécnica del Litoral, Ecuador
[4]Departamento de Infraestructura Geocientífica y Servicios,
Instituto Geológico y Minero de España (IGME), Spain

ABSTRACT
Water sanitation is one of the biggest challenges nowadays. The sixth sustainable development goal of the 2030 agenda of the United Nations focuses on the importance of clean water and sanitation worldwide. El Rosario parish (Ecuador) was founded in 1991 and has never had a sanitary sewer system and a wastewater treatment system since. The households of the 47 hectares directly discharge domestic wastewater to a stream which flows into Daule river. This river is the primary source of drinking water for more than 4 million people that live in the province of Guayas. It implies problems with health, hygiene, and significant environmental impacts. This paper aims to develop a technical proposal for the design of sewerage and wastewater treatment, through technical analysis of the area of study, mitigating environmental and health impacts. The methodological process includes: (i) acquiring relevant data like current population, information related to the drinking water system and water supply, topography, weather, and water resources; (ii) layout of primary, secondary, and tertiary branches of the sanitary sewage system; (iii) calculation of the direction of nearby water resources' drainage that lead the layout of primary and secondary branches of the storm sewage system; and (iv) analysis for the selection of the wastewater treatment system, capacity, location, and design. As a result, this work proposes the design of stabilization ponds implanted in 14.2 ha with 26 km of sanitary and storm sewer networks in a separative sewage system. In addition, the execution of the proposal is estimated for 9.5 months for a value of approximately USD 2,157,100.35. This solution will mitigate the pollution of the Daule river and improve the life quality of the inhabitants of the rural community.
Keywords: sewage, wastewater treatment system, water quality, sanitation, environmental impact.

1 INTRODUCTION

Water is a main natural resource which requires a previous treatment before its use by humans. However, water needs a treatment after its use and prior its discharge to water bodies [1]. In general, domestic wastewater could be of two types: grey water and black water. Grey water refers to wastewaters from kitchens, sinks and showers. Black water includes wastewaters from toilets. It is important to be able to distinguish the types of wastewater for its appropriate treatment and subsequent reuse [2]. To take advantage of the water resource and preserve it, it is governments' duty to build waterworks like sanitation and depuration systems [3] (e.g. water volumen, surface characteristic of the terrain, hydrogeology).

The magnitude growth of human activities generates a negative impact on the environment [4]. This impact includes big amounts of wastewater that end into water bodies around the world [5]. In general, wastewaters are treated but in some cases the appropriate measurments are not followed. This situation is more common in commuties with low income, like rural zones. In this kind of zones, water bodies like streams are turned into sewerage systems and garbage dumps turning these places into main disease spread points

[6]. An ideal situation is a sewerage system which recolects and transports wastewater to a treatment plant before its discharge.

The absence of a sewage and wastewater treatment system can have serious consequences in people's health because a poor sanitation environment is generated. Among the most common water diseases are dengue, cholera, and dysentery. But, in extreme cases, critical rates of malnutrition can occur like in Africa and Asia; where they have a rate of 23.8% and 46%, respectively [7]. Water pollution diseases are very common. In 2017, approximately 220 million people required treatment for schistosomiasis, a disease that originates from contaminated water. Another related disease is diarrhea where 842,000 people die each year, and which 43% are children under 5 years old and their death could be prevented considering their condition at risk. Japanese encephalitis, where 25% of clinical cases die, while 35% are left with permanent brain sequelae, and many others such as hepatitis A, schistosomiasis, and fluorosis, which are easily transmitted in an unsanitary environment [8].

Wastewater treatment plants management is an important problem. Not only for industries and municipal governments, but especially for environmental management and its impacts. If wastewater is managed improperly, then raw water supplies for transformation into drinking water will suffer. If wastewater is discharged without the necessary treatment, the environmental quality will be affected. To solve this, wastewater treatment focuses on reducing the amount of carbonaceous materials (organic; determined predominantly as biochemical oxygen demand) and, when sensitive waters are involved, nitrogen (N) and phosphorus (P) compounds, before from being discharged to the receiving systems [9]. The presence of these materials in high concentrations can have deleterious effects on the levels of dissolved oxygen (O_2) concentration, the trophic state and, ultimately, the well-being of the fauna and flora in the water.

There is a great disparity between developed and developing regions, as well as between rural and urban areas, especially in Latin America, South Asia and Oceania. In this zones, the values indicate that 7 out of 10 people who do not have access to basic sanitation, they live in rural areas [10]. Ecuador has indicated that 90% of the population has access to sanitary sewage systems in urban areas. But, Guayas province has a coverage of 61.7%, and the capital Guayaquil has coverage 73% of sanitary sewers. El Empalme is among the lowest in terms of sewerage coverage with a range of 10%–38% [11]. The parish of El Rosario has a excreta evacuation system which consists in putting domestic wastewater in a natural estuary that flows into the Daule River. The untreated discharge into the Daule River becomes a potential public health problem for the Guayas province, since this river network is a source of drinking water and irrigation for crops, which could cause gastrointestinal diseases [12]. One of the most common diseases in the study area is malnutrition, which occurs in most cases due to unsanitary conditions, encouraging the proliferation of different types of parasites. El Rosario has a chronic malnutrition rate of 43.8%, ranking among the highest percentages at the cantonal level and exceeding the global malnutrition rate of 16% [13], [14].

Regarding the techniques and methods described in the literature to address the issue of wastewater, the sewerage system includes a group of pipes added to typically reinforced concrete structures that help to correctly distribute the intervened flow. This system is strategically connected to collect and transport wastewater and/or rainwater, depending on its type and function [15]. Stabilization ponds are a simple method for treating wastewater, since it does not require sophisticated equipment, nor many resources for its construction, operation and maintenance. For the reasons stated, this type of depuration method is ideal to be implemented in small communities that do not count with many economic resources. In addition, stabilization ponds are very efficient in removing organic matter [16], which is a pollutant that characterizes domestic wastewater [17].

1.1 Objective

This document's objective is "to develop the design of the sewerage system and wastewater treatment system of the El Rosario parish, in such a way as to improve the quality of life of its inhabitants, and guarantee the water resource for human consumption, through the technical analysis of the study area and sustainability factors, mitigating the contamination of the parish and the Daule river". A simplified separative system is proposed as a sanitary sewerage and storm sewer system, added to a purification system that contains a grit tank and a set of stabilization ponds. This system has a bypass so that only the first rains of the winter season can enter the purification system, after this, the rainwater can be discharged directly to the Daule river.

1.2 General characteristics of the study area

The study area has a topography with elevations that vary from 25 to 75 masl, consequently it has micro-basins within its geographical limit that will allow the natural discharge of runoff. It is made up of 3 micro-basins that discharge to the Daule River at different points. Its current population was obtained by counting properties and statistical methods based on the last national census and on the average number of inhabitants per household in rural areas. Having a population for the year 2020 of 3,582 people.

The "El Rosario" parish of the El Empalme canton lacks a sewage and wastewater treatment system. The sanitary waste from each house is dumped into the natural stream that runs through the entire parish and finally flows into the Daule River. This action causes an unhealthy environment for the inhabitants of the parish, where the contamination of the estuary is added to the direct contamination of the river into which it flows. The river is a source of raw water to make the water of Greater Guayaquil drinkable (Guayaquil, Daule and Samborondón) affecting a total of more than 4,000,000 inhabitants.

2 METHODOLOGY

To develop the sewerage design applied to the parish head of El Rosario, a methodological sequence adapted to the case study was defined and applied. The steps followed were those indicated in Fig. 1.

In phase 1 the collection of relevant data such as: current population, information related to the provision of drinking water, topography, climate, and nearby micro-basins of the area is approached to obtain the design population and its flows. Since the study area does not currently record census data, the current population was determined by counting the properties and multiplying by a factor of 6, which is the average number of inhabitants per household in the area. With this information, we proceeded to design the sanitary sewage system for a period of 20 years, calculating the future population through arithmetic, geometric and exponential methods. Finally, for the preliminary information section, the hydrographic network of the area was determined, which allowed us to know the point of natural discharge of the waters and thus determine the drainage scheme of the storm sewer system.

In phase 2 the design of the sanitary sewer system is proposed. Specifically, the definition of the layout and design of the primary, secondary and tertiary routes for the discharge of

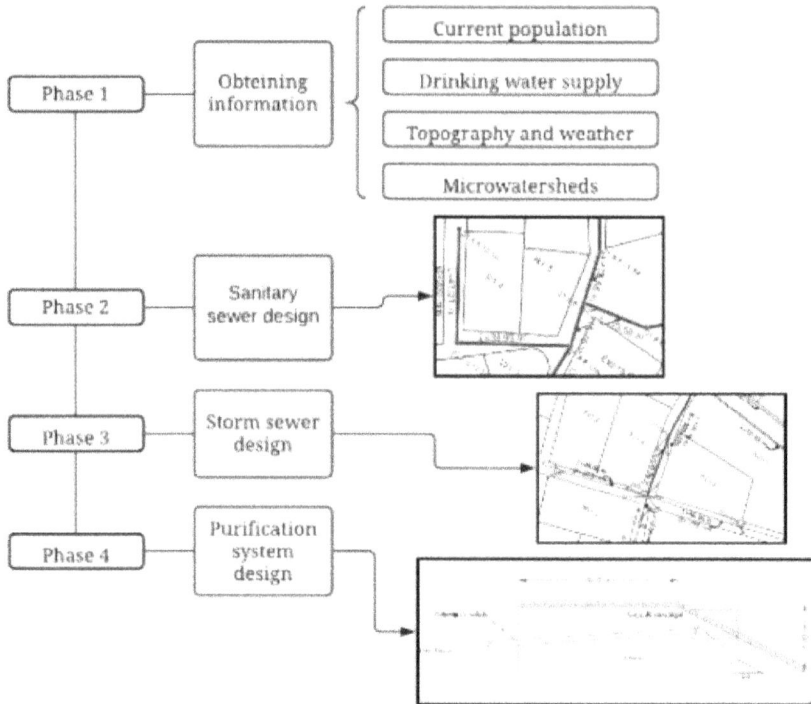

Figure 1: Methodological scheme.

domestic water. The first step was to trace the route of the collectors so that they follow the natural shape of the terrain and avoid the implementation of pumping stations. In this step, revision wells were located at each change of direction and at a maximum separation of 100 m. Then, the contributing and expansion areas were distributed for the different revision wells and the tributary and accumulated areas were calculated. Next, the domestic, institutional, infiltration flows with a factor of 0.15 l/s/ha and illegal water with a factor of 0.2 l/s/ha were calculated. The supply of drinking water taken was 100 l/hab per day, and a return coefficient of 0.8 was used for its transformation into wastewater. From the flow rates obtained, the slopes and diameters were established in accordance with the commercial diameters available in the country. Errors were corrected by means of hydraulic relations and the parameters were verified to check if the slopes, diameters, speeds, and tractive force, within the pipes, are adequate to allow the rapid evacuation of the sewage, without producing sedimentation. or gas production, and preventing erosion by high speeds within the pipes. As a check, you have to ensure:

1. $T > 0.12$ (guarantee traction)
2. $V_{min} \geq 0.45$ m/s (guarantee self-cleaning)
3. $V_{máx} \leq 4.5$ m/s (avoid erosion)

If these parameters are not fulfilled, changes in diameters or slopes must be done. Finally, the start and end terrain, crown, water line, and invert elevations were calculated for each section. These were sketched for a better on-site view and to verify its functionality (see Figs 2 and 3).

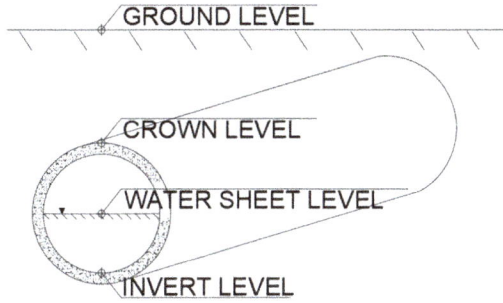

Figure 2: Location of the different dimensions in the collector cross section.

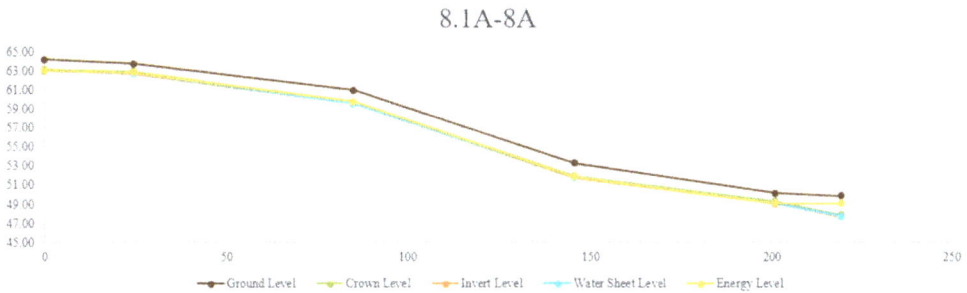

Figure 3: Diagram of dimensions in section 8.1A-8A.

3 RESULTS

The results for sanitary sewer, storm sewer and treatment plant are shown in Tables 1–4.

4 ANALYSIS OF RESULTS

The population estimate is as expected for these rural communities. The water supply is a bit high, taking into consideration that for rural areas it is usually between 50–60 L/person per day, this could be explained by the tropical climate or by losses in the drinking water pipelines.

Table 1: Data results about population, topography and drinking water supply.

Population	
Current (2020)	Future (2040)
3,582 hab	4,846 hab
Topography	
Min.	Max.
15 msal	75 msal
Drinking water supply	
100 L/hab/day	

Table 2: Data results from sanitary and storm sewerage systems.

Sanitary sewerage system		
Length	19.20	km
Diameter	200–250	mm
Slope	0.005–0.4	m/m
Storm sewerage system		
Length	6.80	km
Diameter	250–700	mm
Slope	0.005–0.4	m/m

Table 3: Screen and grit results.

Screen		
Dimensions	30×15	mm
Grit		
Volume	1.8	m^3

Table 4: Ponds for clean water results.

Facultative ponds		
Quantity	4	u
Width	47	m
Length	90	m
Deep	1.5	m
Retention (dry season)	11	days
Retention (wet season)	8	days
Removal (dry season)	79.52	%
Removal (wet season)	73.85	%
Incoming flow	2,280.96	m^3/day
Maturation ponds		
Quantity	2	u
Width	71	m
Length	140	m
Deep	1.3	m
Retention (dry season)	12	days
Retention (wet season)	9	days
Removal (dry season)	90.50	%
Removal (wet season)	86.32	%
Incoming flow	1,569.51	m^3/day

The acquired topography is quite consistent with the evident unevenness in the parish. The micro-basins found agree with the natural exhaust route of the estuary that crosses the parish.

The sanitary sewer system has pipe dimensions and slopes that follow the provisions of the CPE INEN 5 PART 9-1 standard, guaranteeing that there will be no sedimentation or

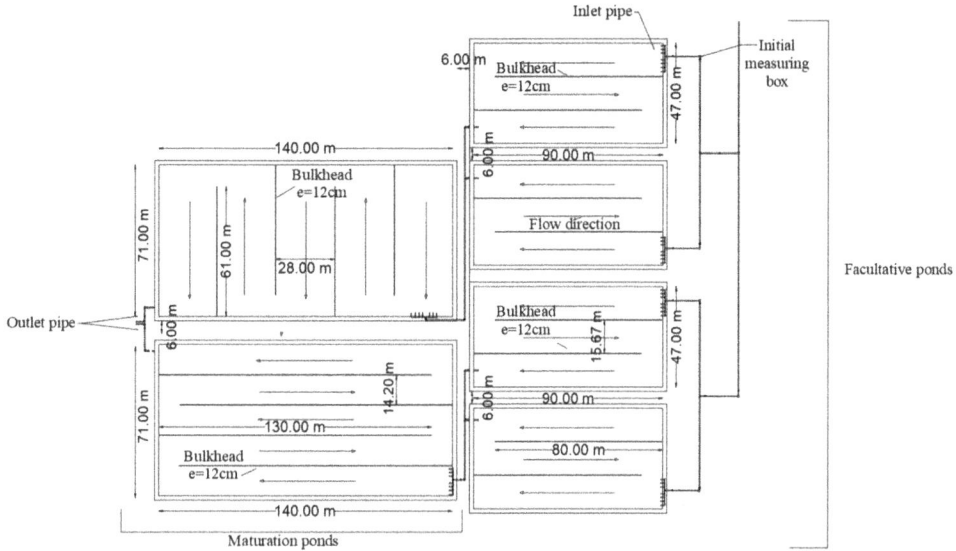

Figure 4: Diagram of depuration ponds.

erosion within the collectors. On the other hand, the storm sewer was drawn only in the paved areas in order to avoid obstructions by mud, stones and sticks; for them it has a shorter total length.

Because the location of the treatment plant is at a higher elevation, due to the lack of space in the initial site, it has been decided to implement a pumping station that works from two centrifugal pumps of 20 HP. The treatment plant will comprise sieves 30 mm × 15 mm, for removing solid particles larger these openings in the area.

When calculating the volume of the facultative lagoons, it was obteined a retention time of 11 days in dry weather and 8 days in rainy season was obtained. This implies that the design process is correct with the values recommended by Crites and Tchobanoglous [18], which stands out a range between 7.5 and 22.5 days.

The configuration of the lagoon system depends on the designer's criteria, but is based on economic and biological aspects of the place of implantation, as well as the facilities for operation and maintenance (OPEX). Taking this consideration, the system will consist of two design lines. Additionally, the lagoon system is considered the most efficient solution in rural areas where low-skilled labor is required for OPEX and where the cost of land is not significant.

For the 2-line design system, the configuration of each line comprises: two facultative ponds in parallel connected to a maturation pond in series. This completes the retention times for each one and removes the amount of pollutants acceptable for discharge, of 100 mgO$_2$/L for BOD5 and 1,000 MPN/100ml for coliforms, in accordance with local regulations. However, in this case, they went further, the removal of fecal coliforms and BOD5 at the end of the process indicate an efficiency of 91%. It was obtained a final concentration of BOD5 of 17.11 mgO$_2$/L in dry weather, and an efficiency of 86% in the rainy season, with a final BOD5 concentration of 24.62 mg/L, which allows the reuse of the waters in possible initiatives for resource recovery.

5 CONCLUSIONS AND RECOMMENDATIONS

From the information obtained in phase 1, it was possible to carry out a sewer network design where the topography helped it to work under gravity in its task of collecting and transporting household waste, however, because the treatment plant It is in an area located above the flood level, a pumping point made up of two 20 HP centrifugal pumps was required.

Diameters were calculated in such a way that the most economical is obtained so that the sewer design is efficient with respect to the common problems that occur in pipes, such as sedimentation or erosion. Thus, system failures or collapses are avoided, and maintenance concurrency is reduced. This design phase contemplates an approximate cost of USD 624,836.60.

For the design of the rainwater network, a design that requires the least possible budget and maintenance expenses was defined, giving an estimate of USD 559,776.65. In the same way as the sanitary sewer network, economic diameters were chosen for the pipes. It was also verified that the speeds do not allow sedimentation and the self-cleaning of the network is guaranteed.

A stabilization pond system was selected as a purification system because this type of system is ideal for a developing rural population. Furthermore, this water purification proposal presents low maintenance and operation costs. This lagoon system has an estimated referential budget of USD 393,954.77.

To improve management of the construction process of the sewerage network, it is essential to carry out a study of the location of existing drinking water supply pipes. Thus complying with standards of distance between sanitary and drinking water pipes. In this way, a design that does not interfere with the existing supply network can be made.

When designing lagoons for a rural area it is important to take into account the availability of a large area for the location of the lagoons. It is recommended, not only that the area is large, but that it is located at a prudent distance from the residential area to avoid the transport of bad odors to inhabited areas. Also consider aspects of the climatic conditions of the site, observe which is the predominant orientation of the wind and arrange the lagoons in that sense.

The design of the sewerage and wastewater treatment system of the El Rosario parish allows the GAD to have an estimate of the cost of the system, to foresee its financing and construction. The projected budget for this project is USD 2,157,100.35, broken down as follows: USD 624,836.60 corresponds to the sanitary sewer, USD 559,776.65 the storm sewer, USD 966,990.66 the purification system and USD 5,496.43 due to the expenses generated by environmental impact. The total work will have an estimated duration for construction of 9.5 months.

Due to the mobilization restrictions because of COVID, the topography data could not be verified, therefore, as a recommendation, it is indicated to adjust these values and the corresponding trace, when the COVID-19 pandemic ends. Try to carry out corresponding sampling at the outlet of the household guides to characterize the domestic discharge and correct the pollutant load with which it reaches the purification system, however, it is considered that the value of 180 mgO$_2$/L for the BDO5 is correct, depending on of the results found in other nearby sites.

REFERENCES

[1] Kataržytė, M. et al., Fecal contamination in shallow temperate estuarine lagoon: Source of the pollution and environmental factors. *Marine Pollution Bulletin*, **133**, pp. 762–772, 2018. https://doi.org/10.1016/j.marpolbul.2018.06.022.

[2]   Vakil, K.A., Sharma, M.K., Bhatia, A., Kazmi, A.A. & Sarkar, S., Characterization of greywater in an Indian middle-class household and investigation of physicochemical treatment using electrocoagulation. *Separation and Purification Technology*, 2014.

[3]   Martín Vide, J.P., *Ingeniería de ríos*, ed. Segunda, Edicions de la UPC: Barcelona, 2007. (In Spanish.)

[4]   Corvalán, C., Hales, S. & McMichael, A., *Ecosistemas y Bienestar Humano*, OMS, 2005. (In Spanish.)

[5]   Pereda, O. et al., Impact of wastewater effluent pollution on stream functioning: A whole-ecosystem manipulation experiment. *Environmental Pollution*, **258**, 2020. https://doi.org/10.1016/j.envpol.2019.113719.

[6]   Dave, M. & Castillo, M., Bioindicadores de qualidade de água como ferramenta em estudios de impacto ambiental. *Revista da FAPAM*, **2**(1), 2003.

[7]   Cueto, M., Cólera y dengue en Lima al final del siglo XX y comienzos del XXI : salud y la cultura de la sobrevivencia. *Historia Social Urbana*, Espacios y Flujos, pp. 253–272, 2009. (In Spanish.)

[8]   Organizacion Mundial de la Salud & Fondo de las Naciones Unidas para la Infancia (UNICEF). Agua y salud, 2019. https://www.who.int/es/news-room/fact-sheets/detail/drinking-water#:~:text=El agua contaminada y el,fiebre tifoidea y la poliomielitis. (In Spanish.)

[9]   Mohsenpour, S.F., Hennige, S., Willoughby, N., Adeloye, A. & Gutierrez, T., Integrating micro-algae into wastewater treatment: A review. *Science of the Total Environment*, **752**, 142168, 2021. https://doi.org/10.1016/j.scitotenv.2020.142168.

[10] Senante, M.M., Sancho, F.H. & Garrido, R.S., Estado actual y evolución del saneamiento y la depuración de aguas residuales en el contexto nacional e internacional. *Anales de Geografía de La Universidad Complutense*, **32**(1), pp. 69–89, 2012. https://doi.org/10.5209/rev-AGUC.2012.v32.n1.39309. (In Spanish.)

[11] Senplades, Agua pocle y alcantarillado para erradicar la pobreza en el Ecuador. **120**, 2014. (In Spanish.)

[12] Valenzuela, S. & Jouravlev, A., *Servicios urbanos de agua potable y alcantarillado en Chile: factores determinantes del desempeño*, Naciones Unidas: Santiago de Chile, 2007. (In Spanish.)

[13] Gobierno Autónomo Descentralizado Parroquial "El Rosario". *Plan de Desarrollo y Ordenamiento Territorial*, **9**(1), pp. 76–99, 2010. (In Spanish.)

[14] Fondo de las Naciones Unidas para la Infancia (UNICEF). Datos y cifras clave sobre nutrición. Improving Child Nutrition: The Achievable Imperative for Global Progress. **140**(4), 2011. http://www.unicef.org/lac/UNICEF_Key_facts_and_figures_on_Nutrition_ESP.pdf. (In Spanish.)

[15] García Rojas, J.L., Evaluación del funcionamiento del sistema de alcantarillado condominal en la Zona R – Huaycán, Ate Vitarte. Universidad César Vallejo, 2018. http://repositorio.ucv.edu.pe/handle/20.500.12692/32303http://repositorio.ucv.edu.pe/handle/20.500.12692/32303.

[16] Cajigas, A., *Ingeniería de Aguas Residuales*, McGraw Hill: Spain, 1995. (In Spanish.)

[17] Vidal, G. & Araya, F., Las Aguas servidas y su depuración en zonas rurales: Situación actual y desafíos. 2014. http://www.eula.cl/giba/wp-content/uploads/2017/09/las-aguas-servidas-y-su-depuracion-en-zonas-rurales-situacion-actual-y-desafios.pdf. (In Spanish.)

[18] Crites, R. & Tchobanoglous, G., *Small and descentralized wastewater management systems*, 1st edn. Boston, MA: McGraw Hill, 1998.

REDUCTION OF CARBON EMISSIONS IN A MEDITERRANEAN URBAN WASTEWATER TREATMENT PLANT

MANUELA MOREIRA DA SILVA[1,2], LUÍS CRISTOVÃO[1], DUARTE MARINHO[1],
EDUARDO ESTEVES[1,3], GIL FRAQUEZA[1,3] & ANTÓNIO MARTINS[4]
[1]Universidade do Algarve, Instituto Superior de Engenharia, Portugal
[2]CIMA – Centre for Marine and Environmental Research, Portugal
[3]CCMAR – Centre of Marine Sciences, Portugal
[4]Águas do Algarve, S.A., Águas de Portugal Group, Portugal

ABSTRACT

In the last few decades, with the rapid growth of population, and more than half of them living in cities, the urban wastewater treatment has become a big challenge that consumes many resources, namely energy. In a climate change scenario, the Mediterranean region is facing more frequent water scarcity periods, and urban water reuse can be a solution, at least for supplying some non-potable water uses. In this context, the performance of urban wastewater treatment plants (WWTP) is of utter importance, to produce environmentally safe treated water while reducing energy consumption and carbon emissions (CE). Activated sludge is the biological process most widely used in wastewater treatment and requires aeration systems in order to promote the oxidation of organic matter and ammonia. It is known that the energy consumed in the aeration processes is of major importance for the global WWTP CE. This study was carried out in a WWTP in southern Portugal, wherein an aeration control system that responds in real time to ammonia and nitrate concentrations was tested. The system is set to optimize the duration of the aerated and non-aerated periods, for nitrification and denitrification. During the experimental period, BOD, COD, *Escherichia coli*, TN and TP were monitored in the treated effluent, in order to verify the quality standards that allow its reuse. The aeration control system contributed to a decrease of about 13% of the specific energy consumption, when compared with the corresponding period in previous years, representing a CE reduction of about 1.2 t CO_2 eq, during the experimental period. The treated effluent maintained its high quality standards and can be used, for example, in agricultural irrigation of local crops. Aeration control systems reacting in real time can have an important role to decrease CE of urban WWTPs; however, further research is needed, including more WWTPs and analyzing seasonal variations in energy consumption over the year.

Keywords: urban wastewater treatment, aeration control, energy, carbon emissions, water reuse.

1 INTRODUCTION

Although wastewater treatment plants (WWTP) are of utmost importance to environment and public health, they consume resources, and release emissions with negative environmental impact and contribute to global warming [1]–[3]. To produce environmentally safe treated water, WWTP have appropriate treatment technology including a set of equipment that execute different processes, with high energetic costs [1], [4]–[7], e.g. a lot of energy is required to pump, treat and deliver water. About 20% of the total energy consumption for public utilities by the municipalities is for wastewater treatment [3], [8], [9], and energetic costs correspond to the WWTP main operation cost [2], [4], [5].

The activated sludge process (ASP) is very common for urban wastewater treatment, despite the large amount of oxygen it requires for the oxidation of organic matter and ammonia in the biological tanks [5], [9]. The requirements in terms of dissolved oxygen to biological processes are closely related to the effluent criteria, defined by legal requirements. Over time, some studies aimed to optimize the ASP related to aeration control, selecting more efficient aerators or using more efficient aeration processes, in order to decrease the energetic consumption [9]–[11], while meeting the quality requirements of the treated effluent.

In order to improve the sustainability of the urban water cycle, the waste water treatment technologies and the water reuse are of paramount importance [12]. The global carbon emissions (CE) of a WWTP are closely related to greenhouse gas emissions (GHG) associated with material consumptions and energy, namely from all treatment processes, e.g. carbon dioxide emission from electric power consumption in the ASP [13]. According to the Portuguese supplier of electricity [14], during 2016 in Portugal 79.6% of electricity comes from renewable sources (wind, water, renewable cogeneration, urban solid waste and others) and 20.4% from non-renewable (coal, natural gas, fossil fuel, nuclear, and municipal solid waste).

The urban wastewater is a resource suitable for the recovery of valuable materials (e.g. nutrients for agriculture) and clean water, and its reuse is a common practice in several countries [12]–[15]. The water reuse is an important measure to address water scarcity reducing the pressure on natural water resources, in a climate emergency scenario, as we are facing in Mediterranean. Several studies demonstrated that water reuse can save chemicals and electricity needed to pump, treat and distribute water from natural sources [12], [13], [16], however the WWTP flow rate availability, effluent quality (e.g. salinity) and the distance between WWTP location and places of application of reused water influence the magnitude of application [13]. In addition, the water reuse can contribute to the implementation of circular economy practices in water sector, decreasing the corresponding CE.

This study was carried in an urban WWTP in the Mediterranean area and aimed to quantify the impact on CE, of installing an aeration control system (RTC-N/DN System®) adjusted to minimize energy consumption and comply with the quality requirements for water reuse.

2 METHODOLOGY

This study was performed between December 2015 and April 2016 in the Ferreiras WWTP, Albufeira (Algarve, south Portugal), built in 1990 and improved in 2002 to serve 22,160 equivalent inhabitants. This WWTP (Fig. 1) has a preliminary treatment with an automatic screening system (6 mm), and oil and grease removal by mechanical separation. The secondary treatment is composed of two biological treatment lines by ASP, each line consisting of an anoxic tank (140 m³), two aerobic tanks (total volume 2.153 m³) with aeration in sequence, and a circular decanter. After secondary sedimentation, the disinfection is carried out through a UV disinfection system and before the discharge into the Albufeira Stream, the effluent passes through a final maturation lagoon to additional UV disinfection and where some nutrients removal also takes place. In the ASP tested line the aeration in the biological treatment is ensured by 2 turbines (Europelet) with 20.45 kW of power and 1.9 kg O_2/kWh of oxygenation capacity. The RTC-N/DN System® was installed to optimize the length of the aerated and non-aerated periods, for nitrification and denitrification, responding directly to ammonia and nitrate concentrations in the ASP. During the experimental period, the aeration time was adjusted to minimize energy consumption, complying with the requirements defined for Biochemical Oxygen Demand (BOD, mg/L·O_2), Chemical Oxygen Demand (COD, mg/L·O_2), E. Coli (MPN/100 mL), Total Nitrogen (mg/L·N) and Total Phosphorous (mg/L·P), in the treated effluent. All these parameters were quantified weekly, in laboratory according to Standard Methods for the Examination of Water and Wastewater [17]. In real time, continuously we measured the influent flow (m³/h), temperature (°C) with

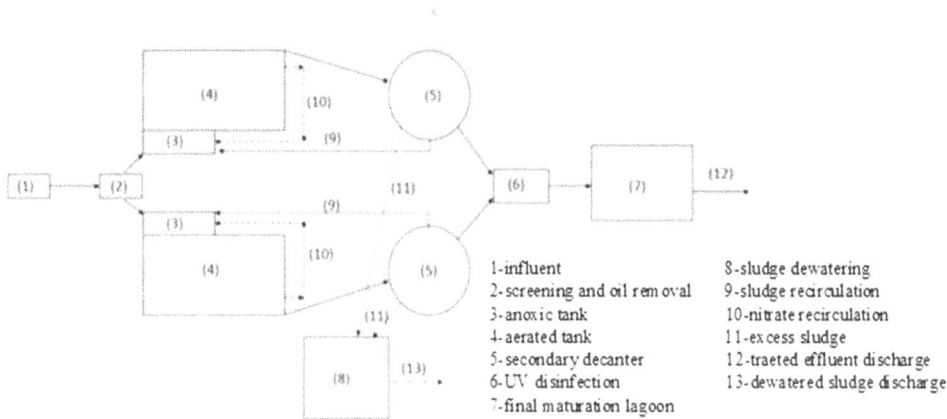

Figure 1: Linear diagram of the Ferreiras WWTP.

the EasyLog EL-USB 2, Dissolved Oxygen (mg/L) with the Hach LDO sensor, and Ammonia (mg/L·N) and Nitrate (mg/L·N) with the combined sensor Hach AN-ISE. All information was saved in a data logger SC 1000. The relationship between parameters was assessed using Pearson correlation (r). The supplier company measured the total energy consumption of the WWTP using an Itrón ACE 6000 meter. It was measured the energy consumption of the turbines during aeration with a mains analyzer Fluke 435 Serie 2.

There are wo types of GHG related to WWTPs, the direct emissions from all processes in the plant, such as CO_2 emission from de ASP, and the indirect GHG emissions associated with the energy consumption [18]. This study focused on indirect GHC emissions related to energy consumption on ASP aeration. The aeration periods were converted into energy consumption (kWh) and CE (kg CO_2 eq), considering 381 g CO_2 eq/kWh [14]. The results of BOD and faecal coliforms in the treated effluent, were compared with the water reuse requirements for agricultural irrigation (Portuguese Law 119/2019).

3 RESULTS AND DISCUSSION

During the experimental period (December 2015 to April 2016), the influent flow ranged between 1,190 m³/day in February and 1,412 m³/day in December, with an average value of 1,286 m³/day. The change evolution (Fig. 2) reflected the pattern of domestic water consumption, with peaks in the early morning and the end of the day, as described in previous works for municipal WWTP [19]. The mean temperature of the influent changed between 17.9°C (at 10:00 h in February.) and 19.1°C (at 18:00 h in April). Two ammonia and nitrate setpoints were tested in the biological reactor in order to evaluate the operation of the control system and the reliability of the probes. On December 2015, at the beginning of the experimental period, the RTC-N/DN System was programmed to start aeration at ammonia concentration of 10 mg·N/L and nitrate concentration of 10 mg·N/L. From February 24 until the end of the experiment (April 30), the target concentrations changed to 5 mg·N/L of ammonia and 5 mg·N/L of nitrate. Fig. 2 shows the change throughout the day during the different months (average and standard deviation) of the influent flow rate, temperature, aeration duration, dissolved oxygen, ammonia and nitrate concentrations. The aeration time throughout the day showed a strong correlation with dissolved oxygen (r = 0.93) and temperature (r = 0.75), while the correlations with ammonia and nitrate concentrations were lower (respectively, r = 0.67 and r = 0.68). Ammonia and temperature showed similar patterns

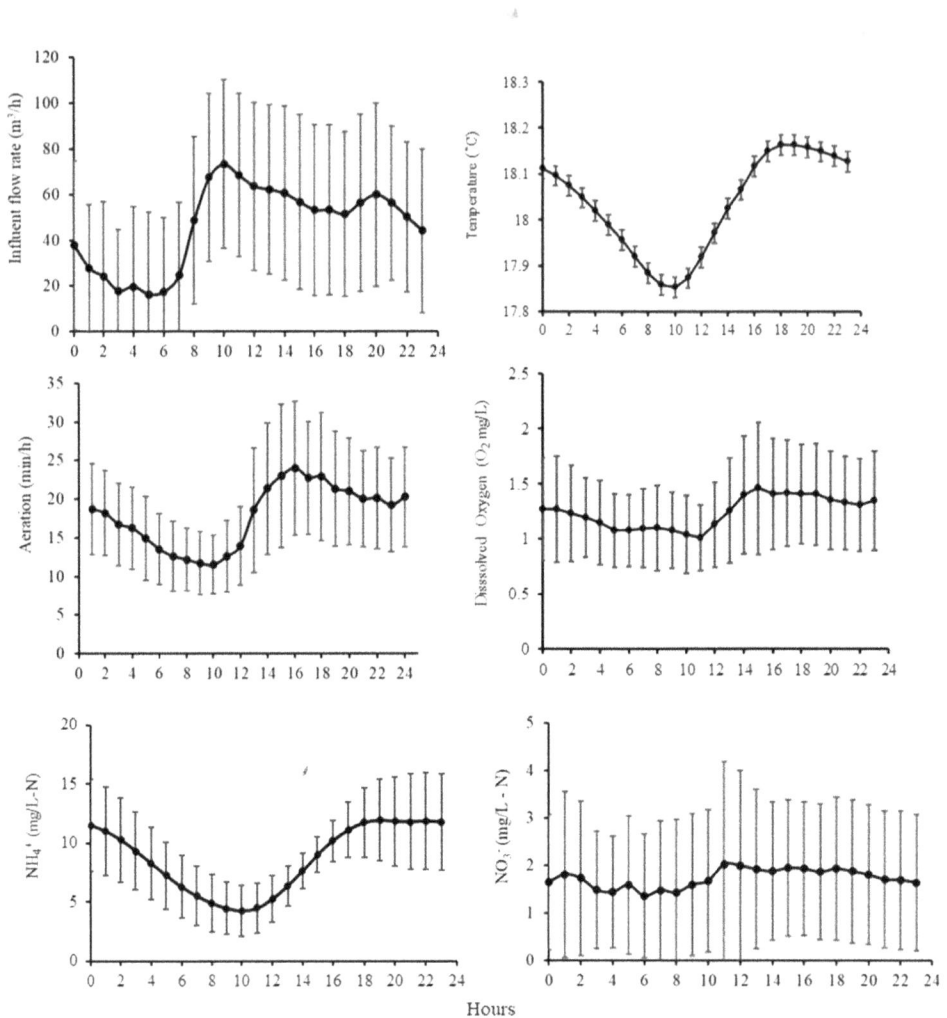

Figure 2:   Daily changes in monitored parameters during the experimental period (average ± standard deviation).

($r = 0.99$), related to the influent flow fluctuation and subsequent oxidation of organic matter, and to the typical variation of the ambient temperature throughout the day.

The results of BOD, COD, and faecal coliforms in the treated effluent (Table 1) met the limits defined in the WWTP Discharge Permit, and the requirements defined for irrigation of crops consumed raw, which grow above the ground, and where the consumable part does not contact water, such as fruit orchards. The water reuse on local fruit tree crops irrigation, such as citrus tree, can function as an important contribution to the water resource´s protection in region, where the water scarcity is a usual problem. Despite the BOD <10 mg/L·O_2, it is expected that treated effluent presents higher organic matter contents than the groundwater usually used for crops irrigation, suggesting that the water reuse can have a positive effect on soils organic carbon and thus on water retention capacity [20]–[22]. Furthermore, the

Table 1: Analytical characterization of the treated effluent in the experimental period, December 2015 to April 2016.

Parameter	Minimum	Average	Maximum	Discharge permit	Water reuse[1]
BOD mg/L·O_2	<10 (QL)	<10 (QL)	<10 (QL)	25	<25
COD mg/L·O_2	21	33	42	125	ND
E. coli MPN/100 mL	42	144	370	2,000	≤1,000
Total nitrogen mg/L·N	3.6	7.1	13.0	ND	ND
Total phosphorus mg/L·P	0.5	1.3	3.6	ND	ND

QL: Quantification limit; ND: Not defined; (1): Portuguese Law 119/2019.

nitrogen and phosphorous contents in the treated effluent, instead of being discharged into the environment promoting eutrophication phenomena, can supply part of plants' nutritional requirements, representing a significant reduction in application of synthetic fertilizers, as reported before in other studies [22].

Regarding the energy consumed in the WWTP, during the experimental period, the aeration consumption represented two thirds of the total energy, ranging from 67% in December to 76% in February (Fig. 3), confirming some previous results obtained for municipal WWTPs with similar treatment technology [9], [23], [24], but higher than reported by other authors [25]. The comparison with the same months on previous years (from 2012 to 2014), shows that during the experimental period, the specific energy consumption in aeration decreased in December, January and February, respectively 40, 10 and 21%, and increased slightly in March (more 8%) and April (more 9%), as shown in Fig. 4. This is due

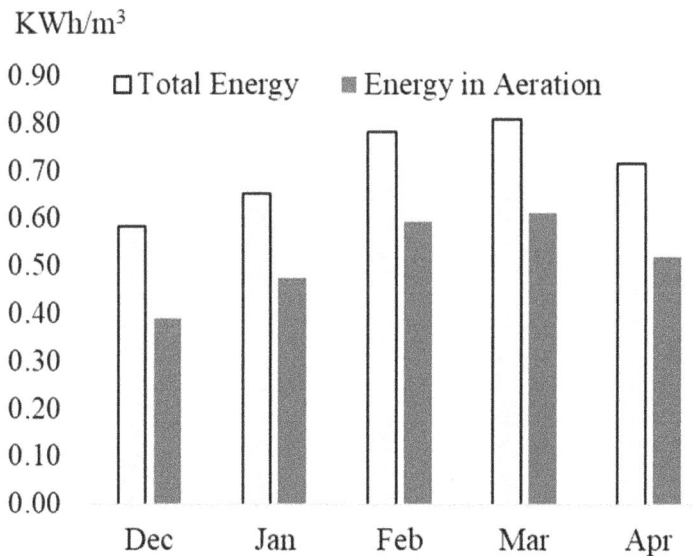

Figure 3: WWTP energy consumption by cubic meter of treated effluent, during the experimental period, December 15 to April 16.

KWh/m^3

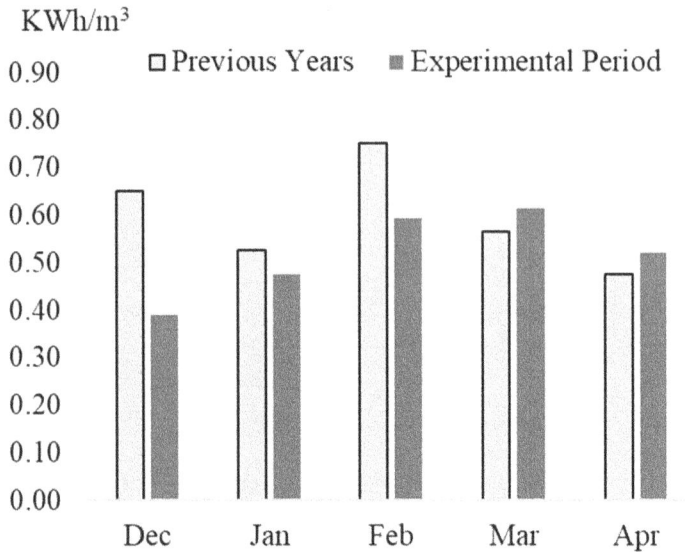

Figure 4: WWTP energy consumption in aeration, by cubic meter of treated effluent in the period 2012–2014.

to the occurrence of a more complete nitrification as the ammonia setpoint in the biological reactor decreased from 10 to 5 mg/L·N, which means more aeration and thus an increase on energy consumption. In average the specific energy consumption in the aeration stage decreased about 13% with the installation of the RTC-N/DN System®. In the experimental period, the specific energy consumption in the aeration stage was 0.52 ± 0.09 kWh/m^3 (average ± standard deviation), which are in line with previous studies for municipal WWTPs from USA and Japan, using conventional ASP [26]. The installation of the RTC-N/DN System® ensured enough oxygen for organic matter oxidation, decreasing the specific CO_2 emissions from 226 to 198 g CO_2 eq/m^3 (Table 2), a reduction of about 13%.

Table 2: Carbon emissions related to energy consumption on ASP aeration.

	Previous years g CO_2 eq/m^3	Experimental period g CO_2 eq/m^3
December	248	148
January	200	181
February	286	226
March	215	234
April	181	198
Average ± standard deviation	226 ± 42	198 ± 35

4 CONCLUSIONS

The WWTPs strongly contribute to the CE of the urban water cycle. The present work showed how energy consumption in a urban WWTP can be optimized, reducing CE, and keeping the quality standards of treated effluent. During the experimental period, the installation of the RTC-N/DN System® saved monthly about 13% of WWTP total electrical energy consumption, which mean a reduction of about 1.2 t CO_2 eq on CE. This reduction is foreseen to be greater in the summer, due to the increase of influent flow related to seasonal tourism in Algarve. The decrease in energy consumption did not affect the treated effluent quality, which can be used for non-potables purposes, as in agricultural irrigation, saving the fertilizer application and reducing the water abstraction from aquatic ecosystems, in accordance to circular economy principles. To improve the sustainability of urban water cycle, regarding the CE related to effluent treatment, further research is needed to optimize the control of the aeration systems, during all the year and including several WWTPs with different technologies.

REFERENCES

[1]   Meneses, M., Concepcíon, H., Vrecko, D. & Vilanova, R., Life cycle assessment as an environmental evaluation tool for control strategies in wastewater treatment plants. *Journal of Cleaner Production*, **107**, pp. 653–661, 2015.

[2]   Mustapha, M.A., Manan, Z.A. & Alwi, S.R.W., A new quantitative overall environmental performance indicator for a wastewater treatment plant. *Journal of Cleaner Production*, **167**, pp. 815–823, 2017.

[3]   Singh, P. & Kansal, A., Energy and GHC accounting for wastewater infrastructure. *Resources, Conservation and Recycling*, **128**, pp. 499–507, 2018.

[4]   Li, W., Li, L. & Qiu, G. Energy consumption and economic cost of typical wastewater treatment system in Shenzhen, China. *Journal of Cleaner Production*, **163**, pp. S374–S378, 2017.

[5]   Markov, Z., Jovanoski, I. & Dimitrovski, D., Clean technologies multi-criteria analysis approach for selection of the most appropriate technology for municipal wastewater treatment. *Journal of Environmental Protection and Ecology*, **18**(1), pp. 289–303, 2017.

[6]   Vasovic, D., Janackovic, G., Malenovic Nikolic, J., Musicki, S. & Markovic, S., Multimodality in the field of resources protection. *Journal of Environmental Protection and Ecology*, **19**(4), pp. 519–1525, 2018.

[7]   Wang, Y., Yuan, L.J., Yan, F. & Zhao, J.Q., Biological nitrogen removal from wastewater via sulphate-reducing anaerobic ammonium oxidation: Advances and issues. *Journal of Environmental Protection and Ecology*, **20**(1), pp. 235–245, 2019.

[8]   Means, E., *Water and Wastewater Industry Energy Efficiency: A Research Roadmap*, AWWA Research Foundation: New York, USA, 2004.

[9]   GIKASP, Towards energy positive wastewater treatment plants. *Journal of Environment Management*, **203**, pp. 621–629, 2017.

[10] Lindberg, C.F. & Carlsson, B., Nonlinear and set-point control of the dissolved oxygen concentration in an activated sludge process. *Water Science and Technology*, **34**(3–4), pp. 135–142, 1996.

[11] Amand, L. & Carlsson, B., Optimal aeration control in a nitrifying activated sludge process. *Water Research*, **46**, pp. 2101–2110, 2012.

[12] Capocelli, M., Prisciandaro, M., Piemonte, V. & Barba, D., A technical-economical approach to promote the water treatment and reuse processes. *Journal of Cleaner Production*, **207**, pp. 85–96, 2019.

[13] Mo, W. & Zhang, Q., Can municipal wastewater treatment systems be carbon neutral? *Journal of Environmental Management*, **112**, pp. 360–367, 2012.

[14] EDP, https://www.edp.pt/origem-energia/. Accessed on: 5 Jul. 2017.

[15] Vergine, P. et al., Closing the water cycle in the agro-industrial sector by reusing treated wastewater for irrigation. *Journal of Cleaner Production*, **164**, pp. 587–596, 2017.

[16] Meneses, M., Pasqualino, J.C. & Castells, F., Environmental assessment of urban wastewater reuse: Treatment alternatives and applications. *Chemosphere*, **81**, pp. 266–272, 2010.

[17] Eaton, A.D., Clesceri, L.S., Rice, E. & Greenberg, A., *Standard Methods for the Examination of Water and Wastewater*, 21st edn, American Public Health Association, American Water Works Association and Water Environmental Federation: Washington, 2005.

[18] Mo, W. & Zhang, Q., Can municipal wastewater treatment systems be carbon neutral? *Journal of Environmental Management*, **112**, pp. 360–367, 2012. DOI: 10.1016/j.jenvman.2012.08.014.

[19] Iman, E.H. & Elnakar, H.Y., Design flow factors for sewerage systems in small arid communities. *Journal of Advanced Research*, **5**(5), pp. 537–542, 2014.

[20] Baldock, J.A. & Nelson, P.N., Soil organic matter. *Handbook of Soil Science*, ed. M.E. Sumner, CRC Press: USA, pp. B25–B84, 2000.

[21] Sparks, D.L., *Environmental Soil Chemistry*, 2nd edn, Academic Press: California, USA, 2003.

[22] Becerra-Castro, C., Lopes, A.R., Vaz-Moreira, I., Silva, E., Manaia, C.M. & Nunes, O.C., Wastewater reuse in irrigation: A microbiological perspective on implications in soil fertility and human and environmental health. *Environment International*, **75**, pp. 117–135, 2015. DOI: 10.1016/j.envint.2014.11.001.

[23] Asadi, A., Verma, A., Yang, K. & Mejabi, B., Wastewater treatment aeration process optimization: A data mining approach. *Journal of Environmental Management*, **203**(2), pp. 630–639, 2017.

[24] Nakkasunchi, S., Hewitt, N.J., Zoppi, C. & Brandoni, C., A review of energy optimization modelling tools for the decarbonisation of wastewater treatment plants. *Journal of Cleaner Production*, **279**, p. 123811, 2021.

[25] William, S.E., *Energy Usage Comparison between Activated Sludge Treatment and Rotating Biological Contactor Treatment of Municipal Wastewater*, Williams and Works, Inc., pp. 1–11. http://www.walker-process.com/pdf/RBC_v_AS_energy_comparison.pdf. Accessed on: 1 Apr. 2020.

[26] Gu, Y. et al., Energy self-sufficient wastewater treatment plants: Feasibilies and challenges. *Energy Procedia*, **105**, pp. 3741–3751, 2017.

ASSESSMENT OF PHYSICOCHEMICAL AND BACTERIOLOGICAL PARAMETERS IN THE SURFACE WATER OF THE JUAN DIAZ RIVER, PANAMA

QUIRIATJARYN M. ORTEGA-SAMANIEGO[1,2], INMACULADA ROMERO[1], MARÍA PACHES[1],
ARTURO DOMINICI[3] & ANDRES FRAÍZ[4]
[1]Research Institute of Water and Environmental Engineering, Universitat Politècnica de València, Spain
[2]Ministerio de Ambiente de Panamá, Panamá
[3]Universidad Marítima Internacional de Panamá, Panamá
[4]Wetlands International, Panama City, Panamá

ABSTRACT
Water pollution represents an obstacle to the development of countries since it affects not only the social-economic component but also biodiversity. Little is documented on the state of water quality of the rivers that flow through Panama City, so it is important to be able to determine the degree of contamination whether of natural or anthropogenic origin, in order to take actions that seek to remediate and increase the resilience of wetland ecosystems. For this study, a database of the Ministry of the Environment of Panama of water quality monitoring during the years 2002–2018 from the Juan Díaz River in the Republic of Panama was used. With these data, a space-temporal analysis was carried out to determine significant differences between the study sites using the Kruskal–Wallis Test and between seasons (dry and wet) by means of the Mann–Whitney U Test, and evaluation of the water quality index (WQI). The results indicate that there are significant differences between sites for the parameters of pH, T (°C), conductivity (mS/m), turbidity (NTU), DO (mg/L), BOD_5 (mg/L), TS (mg/L), SS (mg/L), DS (mg/L), NO_3 (mg/L), PO_4 (mg/L), fecal coliforms (CFU/100 mL), T. coliforms (CFU/100 mL) and there are no significant differences between seasons except for the PO_4 parameter. Analysing the WQI values, all the stations sampled are in the ranges from 17 (highly polluted) to 88 (acceptable).
Keywords: physicochemical parameters, bacteriological parameters, WQI (water quality index), spatio-temporal analysis, Juan Diaz river, Panama.

1 INTRODUCTION

The deterioration of surface water bodies means a great crisis for human development [1]. Therefore, it is essential to investigate the causes and effects on the environment since we depend on this resource for domestic and economic activities [2]. Tools are required to detect the influence of pollutants early and indicate the level of impact [3]. The spatio-temporal analyzes of the physicochemical parameters of the water allow to study the short and long changes of alterations in the nature of the water body [4].

According to the Water Quality Monitoring Report in the Hydrographic Basins of Panama Compendium of Results Years 2002–2008 for the study of the basin, the following anthropogenic pressures were detected: water extraction, poor management of agricultural and livestock production, inappropriate use of soils, urban development near drainage areas, sedimentation resulting from deforestation, floods, contamination by solid waste and industrial pollutants [5]. High concentrations of pollutants and eutrophication processes can cause loss of biodiversity in addition to affecting the quality of life of communities. Therefore, the social, political, and economic component must be considered to reduce and eliminate the environmental impacts in the Juan Diaz River basin.

A spatio-temporal evaluation of physical, chemical, biological parameters and WQI (water quality index) was carried out in order to observe significant differences between study sites and seasons.

WIT Transactions on Ecology and the Environment, Vol 251, © 2021 WIT Press
www.witpress.com, ISSN 1743-3541 (on-line)
doi:10.2495/WS210101

2 MATERIALS AND METHODS

2.1 Study area

The Juan Díaz River is located in the urban area of Panama City specifically in the eastern Pacific of the country [5], being for the most part geographically rugged because it is the last site where the Isthmus of Panama was formed approximately 20 million a year [6]. It is bordered to the north by the Canal Basin 115, to the south by the coastal zone of the Bay of Panama, to the east by the Pacora River basin 146 and to the west by Basin 142 whose main river is the Matasnillo [5]. Within this basin are the district of Panama and San Miguelito, seven townships (Las Cumbres, Tocumen, Belisario Porras, Jose D. Espina, Juan Diaz, Pedregal, and Mateo Iturralde) [7]. Sampling points are shown in Fig. 1.

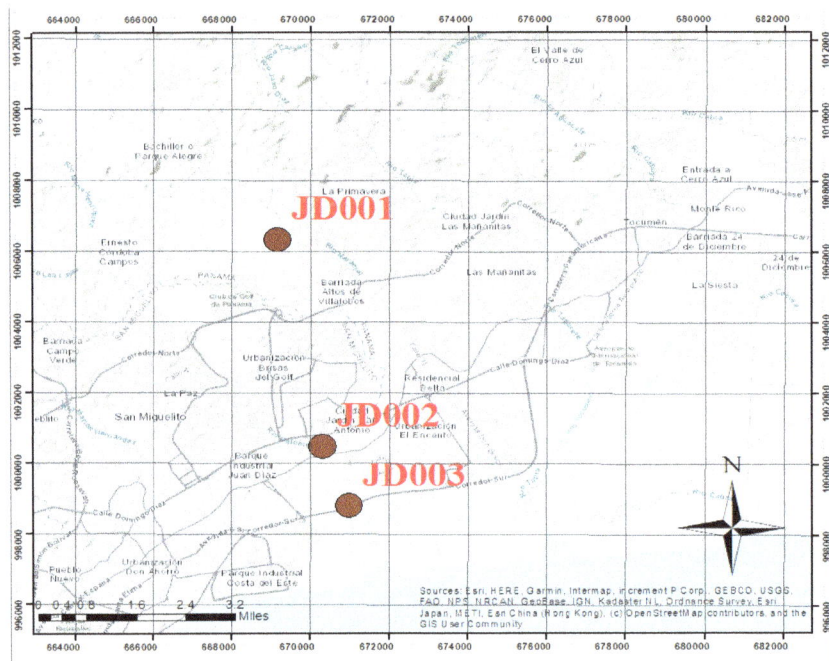

Figure 1: Location of the study sites.

2.2 Analysis of physicochemical and bacteriological parameters

The parameters used in this paper were pH, temperature: T (°C), conductivity (mS/m), turbidity (NTU), dissolved oxygen: OD mg/L, biochemical oxygen demand: BOD_5 (mg/L), total solids: TS (mg/L), suspend solids: SS (mg/L), dissolved solids: DS (mg/L), nitrates: NO_3 (mg/L), phosphates: PO_4(mg/L), fecal coliforms (CFU/100 mL), total coliforms (CFU/100 mL). The data were collected and characterized in the laboratory by technicians from the Ministry of the Environment of Panama following the guidelines of the Standard Methods [8], during the dry and wet season between the years 2002 to 2018 as can be seen in Table 1, the sampling study sites were JD001 (Villalobos Bathing Site), JD002 (Los Pueblos Mall and) JD003 (South Bridge Corridor).

Water and Society VI

Table 1: Descriptive statistics, Kruskal–Wallis and Mann–Whitney Test.

Parameters	N	Minimum	Maximum	Mean	Standard deviation	Kruskal–Wallis Test P(Same)	Mann–Whitney U Test P(Same)
pH	73	6.70	8.40	7.57	0.38	0.00105	0.45866
T(°C)	73	23.50	31.90	27.21	1.84	0.0000182	0.11116
Conductivity (mS/m)	67	6.75	2370.00	102.25	294.09	0.02319	0.75864
Turb (NTU)	66	0.00	246.50	36.26	43.45	0.000000156	0.3077
DO (mg/L)	68	0.00	8.20	4.65	2.49	2.98E-10	0.57186
BOD_5 (mg/L)	67	0.74	348.00	25.62	57.04	9.24E-10	0.73017
TS (mg/L)	73	62.00	1356.00	229.56	189.17	2.97E-11	0.082515
SS (mg/L)	71	1.01	805.00	55.49	108.35	0.0000473	0.32788
DS (mg/L)	71	6.00	1160.21	165.91	141.93	1.27E-09	0.7516
NO_3 (mg/L)	70	0.01	16.60	3.85	3.56	0.000000083	0.34419
PO_4 (mg/L)	60	0.00	5.24	1.37	1.46	0.00000017	0.008335
Fecal coliforms (UFC/100 mL)	48	10.00	8500000.00	543339.89	1756677.28	8.78E-08	0.67237
T. coliforms (UFC/100 mL)	70	215.00	62725100.00	1989812.33	8163926.00	2.37E-10	0.50787
WQI	68	17	88	51.18	20.08	2.45E-10	0.44663

WIT Transactions on Ecology and the Environment, Vol 251, © 2021 WIT Press
www.witpress.com, ISSN 1743-3541 (on-line)

2.3 Data analysis

2.3.1 Spatial analysis

The Kruskal–Wallis Test was applied to determine if there are significant differences for the parameters pH, T (°C), conductivity (mS/m), turb (NTU), DO (mg/L), BOD_5 (mg/L), TS (mg/L), SS (mg/L), DS (mg/L), NO_3 (mg/L), PO_4 (mg/L), fecal coliforms (CFU/100 mL), T. coliforms (CFU/100 mL), between points JD001 (Villalobos Bathing Site), JD002 (Los Pueblos Mall and) JD003 (South Bridge Corridor), using the statistical package PAST 4.01 and SPSS Version 25.

2.3.2 Temporal analysis

The Mann–Whitney U Test was applied to determine if there are significant differences between the dry and wet season of the parameters pH, T (°C), conductivity (mS/m), turb (NTU), DO (mg/L), BOD_5 (mg/L), TS (mg/L), SS (mg/L), DS (mg/L), NO_3 (mg/L), PO_4 (mg/L), fecal coliforms (CFU/100 mL), T. coliforms (CFU/100 mL), using the statistical package PAST 4.01 and SPSS Version 25.

2.3.3 WQI (water quality index)

The WQI (water quality index) was calculated, with a minimum of nine physical–chemical and biological parameters, whose qualification is categorized as highly polluted, polluted, slightly polluted, acceptable, and uncontaminated water quality [9] (Fig. 2).

Rank	WQI	Color
91–100	Uncontaminated	
71–90	Acceptable	
51–70	Slightly polluted	
26–50	Polluted	
0–25	Highly polluted	

$$ICA = \frac{\sum_{i=1}^{n} Li * Wi}{\sum_{i=1}^{n} Wi}$$

Figure 2: Rank of WQI and WQI formula.

3 RESULTS AND DISCUSSION

3.1 Spatial analysis

The variations between the sites for the respective physical–chemical and biological parameters are presented in Figs 3 and 4, the Kruskal–Wallis Test.

Significant differences between points JD001, JD002 and JD003 as can be seen in Table 1. The minimum values for pH, DO are for the JD003 site and the maximum values in T (°C), turbidity, BOD_5, NO_3, PO_4 are for the JD003 site, the maximum values for DO is the JD001 site.

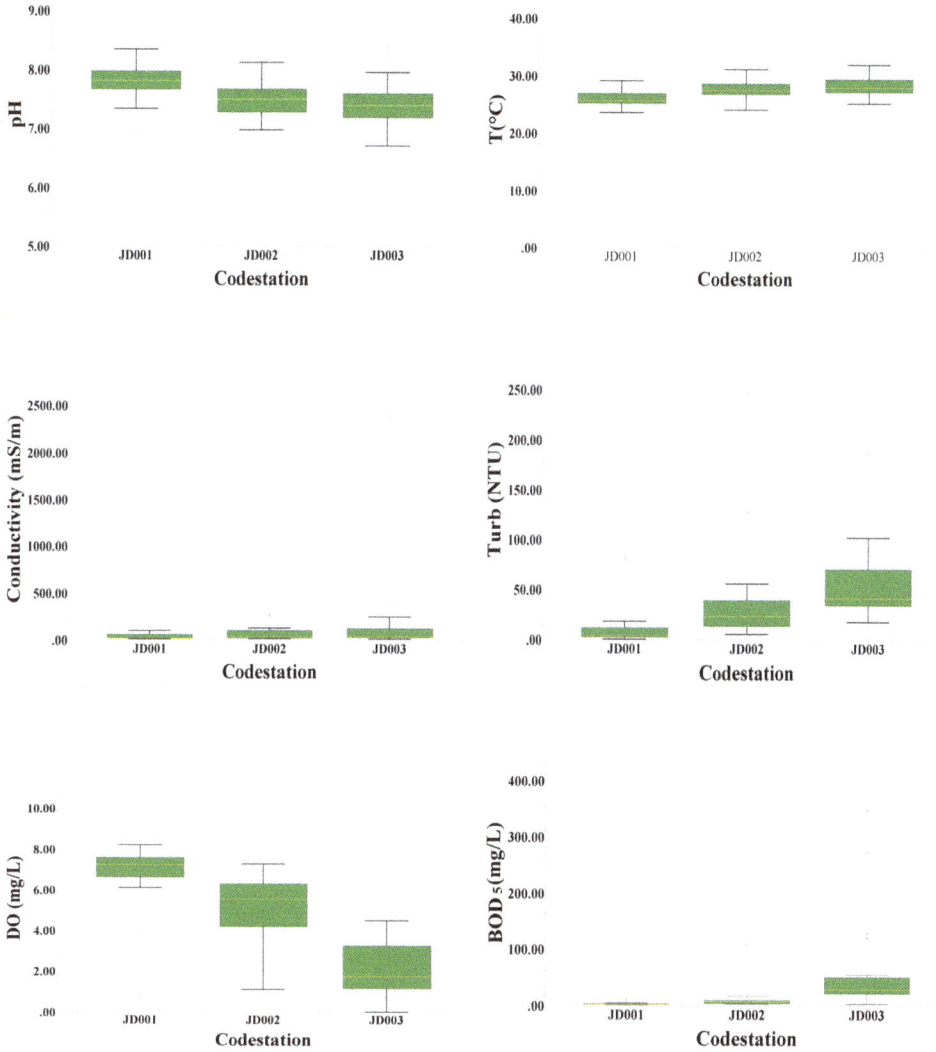

Figure 3: Boxplot of spatial variations of the selected parameters.

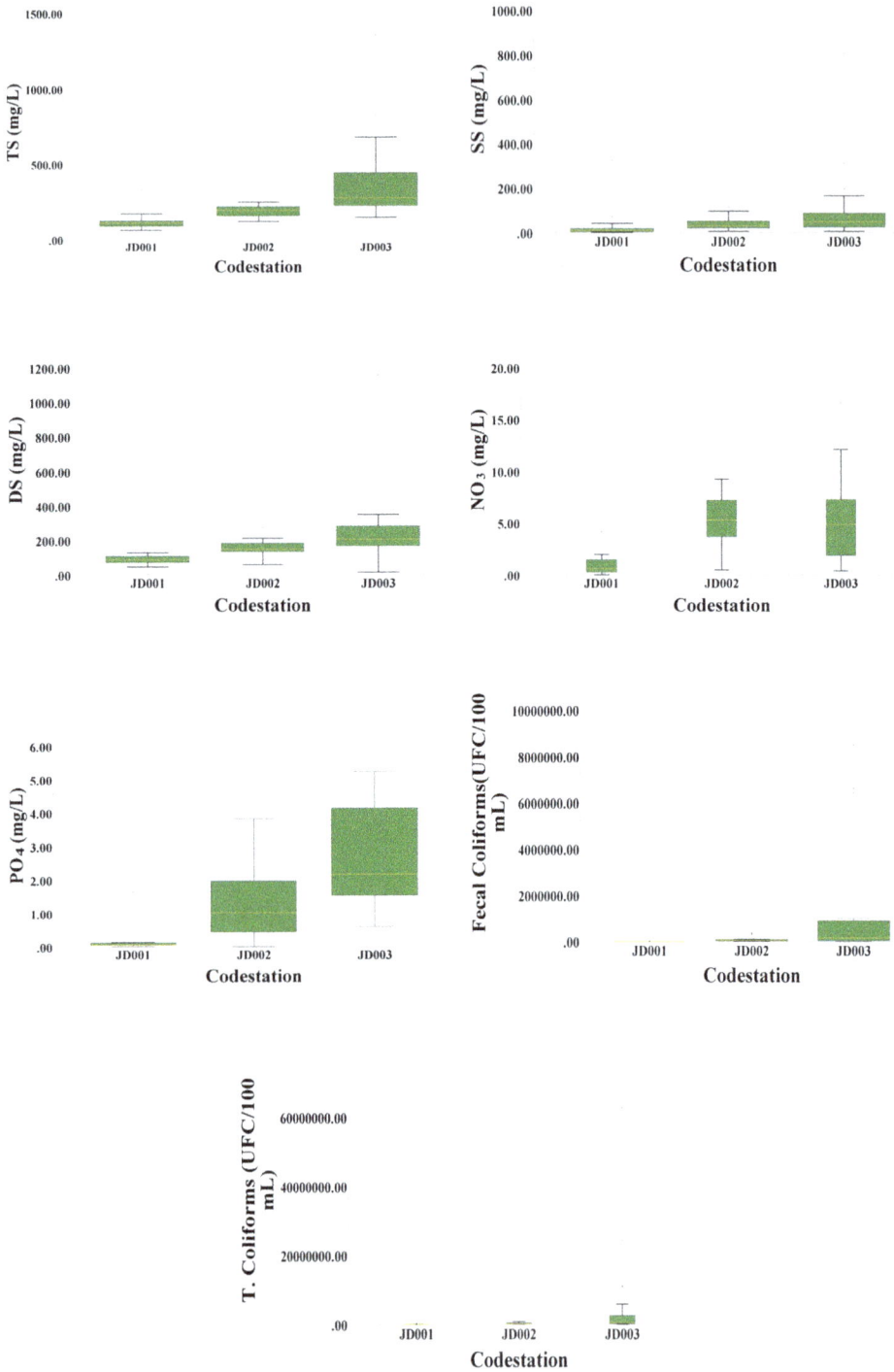

Figure 4: Boxplot of spatial variations of the selected parameters.

3.2 Temporal analysis

The temporal variations for the respective physicochemical and biological parameters are presented in Figs 5 and 6, The Mann–Whitney U Test, as can be seen the results show no differences between the dry and the wet season. The parameter of PO$_4$ (mg/L) may indicate the presence of high intervention of industries that use or produce agrochemicals in the study area and reach the river through the rains.

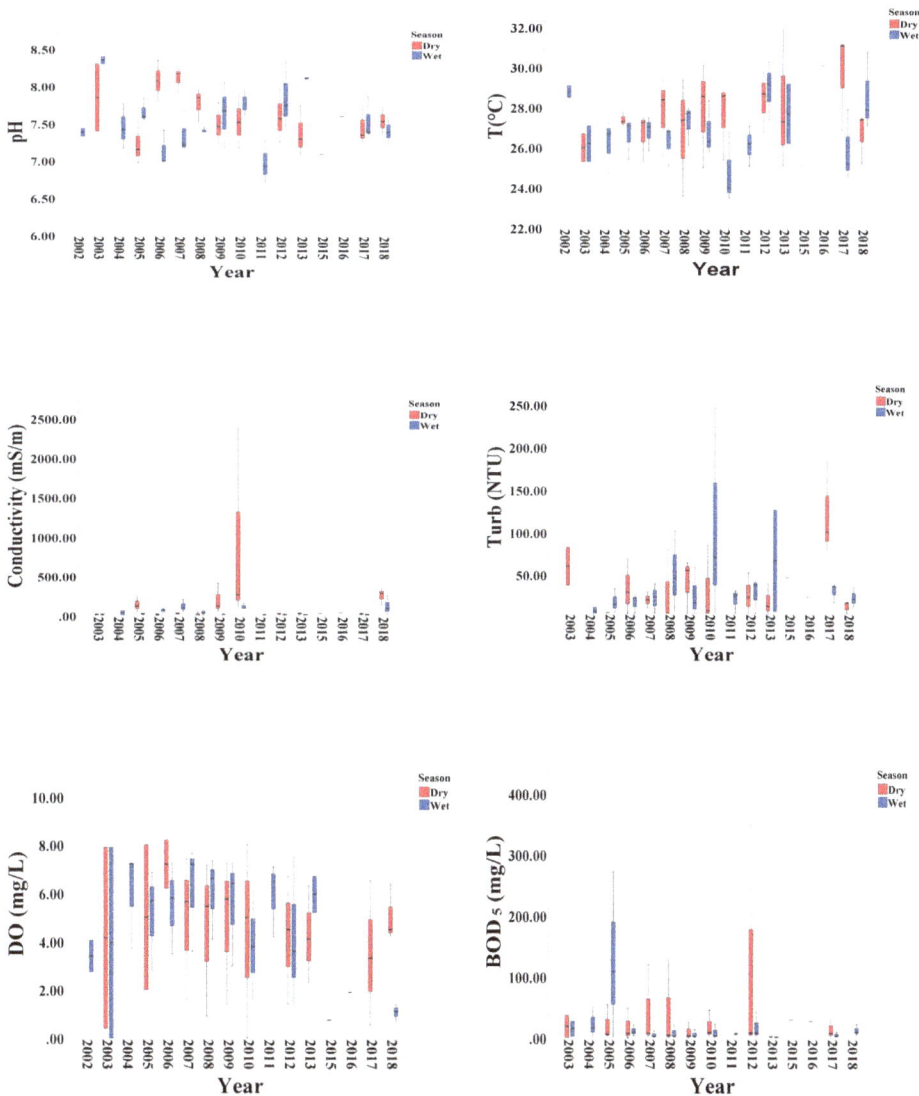

Figure 5: Boxplot of temporal variations of the selected parameters.

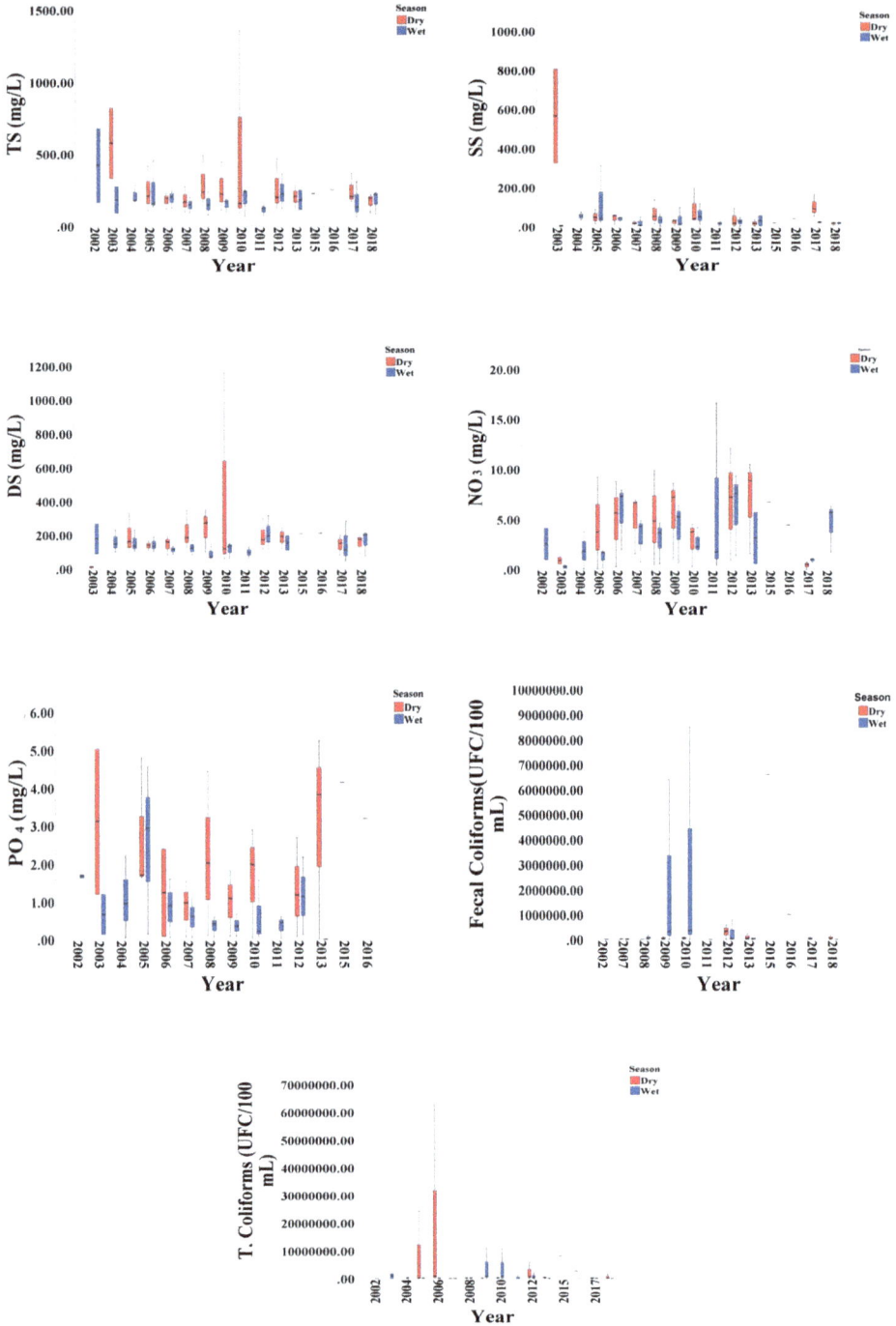

Figure 6: Boxplot of temporal variations of the selected parameters.

3.3 WQI (water quality index)

3.3.1 Spatio-temporal analysis

As can be seen in the graph for site JD001 presented WQI values between 48 and 88 that correspond to the categories of acceptable, little polluted and polluted waters, for site JD002 presented IWQ values between 25 and 68 that correspond to the categories Highly polluted, polluted and little polluted waters, for the JD003 site presented values between 17 and 53, being the category of highly polluted, polluted and little polluted waters. There are no significant differences between dry and wet season (Fig. 7).

The rivers with the greatest deterioration in the country are characterized by a high impact due to urban and industrial development, as is the case of the Juan Diaz river basin [9].

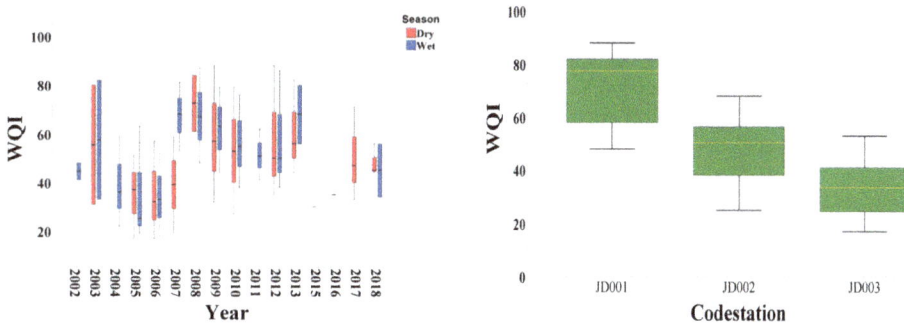

Figure 7: Boxplot of spatio-temporal variations of the selected parameters.

4 CONCLUSION

The results obtained in the Kruskal–Wallis Test indicate that there are significant differences between the study sites JD001 (Villalobos Bathing Site), JD002 (Los Pueblos Mall and) JD003 (South Bridge Corridor) this last point presents the higher concentration of pollutants. The Mann–Whitney U Test indicate that there are no significant differences between the dry and wet seasons except for the parameter PO4.

Analyzing the WQI values, all the stations sampled are in the range from 17 to 88. The highest values of the index appear in the sampling station JD001 (Villalobos Bathing Site) correspond to the categories of acceptable, little polluted and polluted waters, while the sampling station JD002 (Los Pueblos Mall) that correspond to the categories highly polluted, polluted and little polluted waters and JD003 and (South Bridge Corridor) presents the lowest values with the category of highly polluted, polluted and little polluted waters. The pollution levels for the middle and lower zone of the Juan Díaz River are due to anthropogenic environmental impacts: domestic and industrial wastewater discharges and hydromorphological pressures in the river.

ACKNOWLEDGEMENTS

This research was financed by the Scholarship of the Subprogram of Doctoral and Postdoctoral Scholarships of the National Secretariat of Science and Technology (SENACYT) in conjunction with the Institute for the Training and Use of Human Resources (IFARHU). To the Ministry of the Environment of Panama that provided the data for this study. This research is part of the project Environmental Impact of Multiple Stressors in Aquatic Ecosystems of the Metropolitan Area of Panama, financed by SENACYT.

REFERENCES

[1] Tang, T., Cai, Q. & Liu, J., River ecosystem health and its assessment. *J. Appl. Ecol.* **13**(9), pp. 1191–1194, 2002.

[2] Deng, X., Xu, Y., Han, L., Yu, Z., Yang, M. & Pan, G., Assessment of river health based on an improved entropy-based fuzzy matter-element model in the Taihu Plain, China. *Ecol. Indic.*, **57**, pp. 85–95, 2015.

[3] Sasi, S., Rayaroth, M.P., Aravindakumar, C.T. & Aravind, U.K., Identification of surfactants and its correlation with physicochemical parameters at the confluence region of Vembanad Lake in India. *Environ. Sci. Pollut. Res.*, **25**, pp. 20527–20539, 2018. DOI: 10.1007/s11356-017-0563-4.

[4] Wolska, L., Sagajdakow, A., Kuczynska, A. & Namiesnik, J., Application of ecotoxicological studies in integrated environmental monitoring: possibilities and problems. *TrAC Trends Anal. Chem.*, **26**, pp. 332–344, 2007.

[5] ANAM (National Environmental Authority), Report on the Monitoring of Water Quality in the Watersheds of Panama Compendium of Results Years 2002–2008, 2009.

[6] Bacon, C.D., Silvestro, D., Jaramillo, C., Tilston Smith, B., Chakrabartye, P. & Antonelli, A., Biological evidence supports an early and complex emergence of the Isthmus of Panama. *Proc. Natl. Acad. Sci. USA*, **112**(19), pp. 6110–6115, 2015.

[7] Ministry of the Environment (MiAMBIENTE), National Water Security Plan 2015–2050: Water for All. Panama: MiAMBIENTE/CONAGUA, 2016.

[8] APHA (American Public Health Association), Standard Methods For the Examination of Water and Wastewater, 22nd ed., American Public Health Association: Washington, DC, 2012.

[9] Cornejo, A. et al., Diagnosis of the environmental condition of the superficial tributaries of Panama, 2017.

Author index

Abushammala M. F. M. 37
Ahilan S.. 11
Amaya F. 77
Aznar-Crespo P. 65

Berrezueta E. 77
Bréthaut C. 1

Carrión P. 77
Cortés-Borda D. 23
Cristovão L.. 87

Dender L. 77
Dominici A... 95

Escobar-Sierra M. 23
Esteves E. 87
Ezbakhe F.. 1

Fraíz A... 95
Fraqueza G. 87

Jara D. 1

Krivtsov V.. 11

Latif M. F. A. 37

Marinho D. 87
Martin G. 49
Martins A..87
Merchan B. 77
Moreira Da Silva M............................87

O'Donnell E..11
Olcina-Sala Á. 65
Ortega-Samaniego Q. M. 95
Ortiz G. 65

Paches M. 95
Pagano A. 11
Pluchinotta I.......................................11
Polanco J.-A. 23

Qazi W. A. 37

Reisalu G. 49
Rodríguez-Echevarría T......................1
Romero I...95

Tamai K. 57
Torn K. 49

Ullauri P..77

Urban Water Systems & Floods III

Edited by: **S. MAMBRETTI**, *Polytechnic of Milan, Italy and* **D. PROVERBS**, *Birmingham City University, UK*

Flooding is a global phenomenon that claims numerous lives worldwide each year. Apart from the physical damage to buildings, contents and loss of life, which are the most obvious, impacts of floods upon households and other more indirect losses are often overlooked. These indirect and intangible impacts are generally associated with disruption to normal life and longer term health issues. Flooding represents a major barrier to the alleviation of poverty in many parts of the developing world, where vulnerable communities are often exposed to sudden and life-threatening events.

As our cities continue to expand, their urban infrastructures need to be re-evaluated and adapted to new requirements related to the increase in population and the growing areas under urbanization. Topics such as contamination and pollution discharges in urban water bodies, as well as the monitoring of water recycling systems are currently receiving a great deal of attention from researchers and professional engineers working in the water industry. The papers contained in this volume cover these problems and deals with two main urban water topics: water supply networks and urban drainage.

Originating from the 7th International Conference on Flood and Urban Water Management, the included research works include innovative solutions that can help bring about multiple benefits toward achieving integrated flood risk and urban water management strategies and policy.

ISBN: 978-1-78466-379-7   eISBN: 978-1-78466-380-3
Published 2020 / 196pp

www.ingramcontent.com/pod-product-compliance
Lightning Source LLC
Chambersburg PA
CBHW081547220326
41598CB00036B/6592